DR. JOANNE WESCOTT

The Psychological Effects of the Covid-19 Pandemic

About the Author:

FIND OUT HOW DELMARVA COMMUNITY CONSULTANT SERVICES LLC CAN HELP YOU!

Dr. Joanne Wescott is a life learner, servant leader, innovator, and visionary with advanced education. A Licensed Practical Nurse from the commonwealth of Virginia. She then went on to graduate from Delaware State University with a Bachelor of Science in Nursing as a Registered Nurse. She holds a Professional Registered Nurse with Multi-State Licensure, State of Delaware Compact, and a California License. Later obtained a master's degree in Organizational Leadership with a concentration in Mentoring and Coaching. Her most recent achievement was achieving her Doctorate in Strategic Leadership with a concentration in Healthcare leadership. Joanne looks forward to sharing her experience and knowledge in the business of healthcare and professional development arena, shaping an environment that is innovative and forward-thinking. She loves giving back to her community and when she is not working as a consultant, coaching, or speaking and spending quality time with family, she provides community service. She has touched almost every aspect of Nursing throughout her career and has made Utilization Management her specialty. Joanne is an educator and is the CEO of Delmarva CCS where she resides in Delaware.

Please visit Https://delmarvaccs.com for more information, workbooks and guides!

The Psychological Effects of the COVID-19 Pandemic

A view from the socio-cultural context ______________________

By

Dr. Joanne Wescott

Abstract

The world is facing one of the most challenging times in human history, where the coronavirus has made many people lead a life that is not normal. Being forced to live an abnormal life as a new normal brings a lot of psychological discomforts. In this book written by Doctor Joanne Wescott, the psychological effects of the pandemic will be analyzed with reference mainly made to its effects on healthcare providers and the public in general. This is done by analyzing the existing literature reviews on the subject and analyzing the data from the databases of government and concerned agencies, such as the CDC and WHO. Enthusiastic unsettling influence, discouragement, stress, state of mind changes and crabbiness, sleep deprivation, post-horrendous pressure side effects, outrage, and intense fatigue, among those isolated, are seen day to day (Knifton & Inglis, 2020). Misfortune, detachment, loss of pay, and fear set off mental prosperity conditions or increase existing ones (Pfefferbaum & North, 2020). Remarkably, numerous people may be reporting extended levels of alcohol and prescription use, a dozing problem, and pressure. Meanwhile, COVID-19 can incite neurological and mental complexities, similar to absurdity, agitation, and stroke (Pfefferbaum & North, 2020). People with earlier mental, neurological, or substance use issues are furthermore frail against SARS-CoV-2 pollution—they could stand a higher risk of outrageous outcomes and even passing. There is a need to address these challenges and consequences lest they become detrimental to people's lives in a few years to come. These damaging consequences are related to the fact that psychologically distressed individuals cannot be economically productive. Addressing the psychological effects is the initial point of solving the consequences of the effects, both short-term and long-term.

Dedication

This book is dedicated to all the frontline healthcare workers who have worked tirelessly and risked their lives for the sake of the general population and public safety. Also, I believe this book will form a foundational healing point for those the pandemic has directly affected, leading to the loss of their loved ones. Finally, I dedicate this book to all those who have supported me. The pandemic will be contained one day, and we will live happily.

TABLE OF CONTENTS

Preface and Value Proposition

The coronavirus pandemic has dramatically affected the whole world. Its effects are realized in the economy of many countries and all other aspects of life. One of those aspects is the psychological well-being of people. The pandemic has afflicted pain in people, and they tend to find it hard to manage the psychological effects. The book *The Psychological Effects of the COVID-19 Pandemic* is timely and relevant. In these challenging times, many people are undergoing unimaginable mental health issues that require a more profound understanding to be addressed. During the pandemic, many people were forced to live a life they had never known (Kumar & Kumar, 2020). They were forced to adapt to conditions that were not only crazy but were pretty against the social norms or ways of life. For instance, social distancing and wearing masks in public places are the new normal that affect many people (Kumar & Kumar, 2020). Getting infected and being quarantined is a mind-boggling feeling nobody wishes to experience.

Emotional disturbance, depression, stress, mood alterations and irritability, insomnia, post-traumatic stress symptoms, anger, and emotional exhaustion among those quarantined are seen daily (Knifton & Inglis, 2020). Loss, seclusion, loss of pay, and dread set off psychological well-being conditions or intensify existing ones (Pfefferbaum & North, 2020). Notably, many individuals might be confronting expanded degrees of liquor and medication use, a sleeping disorder, and tension. In the interim, COVID-19 can prompt neurological and mental intricacies, like ridiculousness, unsettling, and stroke (Pfefferbaum & North, 2020). Individuals with prior mental, neurological, or substance use problems are additionally more powerless against SARS-CoV-2 contamination. They might stand a higher chance of extreme results and even death.

Understanding these aspects and how to manage them across the affected populations is the motive behind writing this book. The book explains the relationship between the pandemic and the psychological well-being of people. This is accomplished by creating an argument on the pandemic's background, effects, and management schemes. As a nurse, the author has first-hand information on what is needed to address the impact. Through close reading and analysis of this book, one will understand what it takes to manage the associated mental effects of the pandemic and what is required to avoid being affected by these effects.

Moreover, many will be ready and equipped with the measures and approaches needed to overcome the impact of the pandemic. It is within public knowledge that the virus is here to stay, and the prevention and control measures proposed by the concerned agencies may not guarantee 100% chances of safety (Pfefferbaum & North, 2020). Therefore, this book offers robust knowledge on what people require to manage the mental effects of the pandemic.

Chapter 1: Introduction to Mental Health Effects

The Psychological Effects of the COVID-19 Pandemic

In the history of humanity, pandemics and natural disasters are some of the most devastating and mentally disturbing situations. They leave people with much pain, making them question the initial need for existence or creation if the leaders or concerned authorities cannot stop agony, pain, and their associated effects. Not a single pandemic or natural disaster brings positive outcomes. One of the commonly noticed effects of these occurrences is the mental torture of the affected individuals. Many people adapt to a life they have never known whenever pandemics strike. Some people tend to adjust and adapt to conditions that ensure their safety. For instance, the outbreak of the COVID-19 pandemic made many people adjust to the lifestyle that involved "keeping social distancing, avoiding crowded places, wearing masks whenever in public places, and alienating when one suspects he or she has the infection" (Corpuz, 2021, p. e333). Also, during natural calamities like tsunamis, people are encouraged to abandon their homes and move to safer grounds to be saved from catastrophes. Doing all these brings about some psychological discomfort, and the effects may be long-term when the condition goes without adequate and timely address. These changes aim to ensure the population is safe from the effect of pandemics (Corpuz, 2021).

Since the COVID-19 infection was declared a global pandemic, many people have undergone the worst depressing moments, and there are signs that the sad times are not ending soon (Al-Ramadan et al., 2021). The conditions and the pandemic path indicate that people will have to formulate ways and be ready to live with the virus amidst them. Living with the virus within the population is a scary aspect requiring an impromptu address (Corpuz, 2021). Similarly, the HIV/AIDS pandemic got addressed.

This book focuses on the COVID-19 pandemic and its psychological impact or effect on people. In this sense, the term people means both healthcare professionals and ordinary citizens seeking a solution or looking for healthcare workers' (HCWs) services. Over a year into the COVID-19 pandemic, the subject of long COVID, the side effects of the illness that continue for weeks or months, is acquiring unmistakable quality and significance. Specifically compelling are the psychological and mental confusions that can wait in COVID-19 survivors long after they have left the clinic and returned to their homes.

Reports of long-haul weariness and post-traumatic stress disorder issues are expanding in number while aiding in lucidity, taking everything into account (Haseltine, 2021). Some of these neuropsychiatric impacts, like the deficiency of taste and smell, are transient, showing up as the illness advances and dying down once the infection clears (Varatharaj et al., 2020). Nevertheless, others are more relentless and possibly, even long-lasting, like harm from stroke (AlRamadan et al., 2021). However, psychological and mental confusions are imploded into one classification; the more proof arises, the more essential it becomes for the motivations behind care and treatment to recognize the two (Al-Ramadan et al., 2021). Neurological manifestations relate explicitly to the harm the infection causes—regardless of whether straightforwardly or by implication—to the sensory system, while mental indications envelop social, passionate, and mental hindrances (Al-Ramadan et al., 2021).

Neurological COVID-19 indications specifically have been challenging to gather and nail down because of their inconstancy (Varatharaj et al., 2020). Milder structures incorporate weariness, migraines, and the now notorious loss of taste and smell. At the same time, extreme manifestations range from strokes and seizures to encephalopathy (brain disease) and encephalitis (inflammation in the brain) (Varatharaj et al., 2020). Around 2% of patients experience an ischemic stroke, which happens while coagulating blocks the progression of blood to the mind. Moreover, it might happen once in a long while. However, some COVID-19 patients foster either an immune system condition known as GuillainBarré disorder (GBS) or acute transverse myelitis—spinal cord inflammation (Haseltine, 2021). Contrarily, psychological complications are wide-running yet all at once more ubiquitous (Haseltine, 2021). It is usual for patients to end up in the ICU—those with extreme COVID-19 manifestations. However, other sooner or later crises may include experiences of insanity because of the intense pressure of intrusive methodologies like medical procedures and intubation (Haseltine, 2021). Numerous patients who leave the emergency clinic after treatments for primary disease proceed to foster PTSD (Haseltine, 2021). Indeed, even COVID-19 survivors who just had gentle instances of the illness could battle with mind haze, the term credited to the muddled mess of memory and mental loss that can wait months after the infection has cleared.

Many changes have been brought about by the COVID-19 pandemic, especially in how people live and conduct their daily operations and transactions (Jia et al., 2021). The pandemic's disturbing changes include uncertainty of tomorrow, alterations in daily routines, social isolation, and financial constraints following the massive layoff of people and the closure of many institutions (Jia et al., 2021). Notably, it is customary to worry about being sick, the possible timeline of the pandemic, the potential effect the pandemic will have on a job, and what lies ahead. Consequently, many people have gotten the opportunity to keep feeding people incorrect data on the disease (Nelson et al., 2020). Being misinformed about the disease brings more fear and makes life feel out of control without any straightforward approach. It is no news that this pandemic and its associated effects may make one feel stressed, anxious, afraid, sad, and lonely (Haseltine, 2021). These feelings are psychologically disturbing and may plunge one into depression and develop more mental health disorders.

In America, many people, especially the elderly, have recorded high levels of stress, depression, anxiety, and insomnia compared to before the pandemic (Jia et al., 2021). Because of the increase in these mental conditions, many people have resolved to use alcohol and hard drugs, hoping these would help them overcome the fears of the pandemic (Jia et al., 2021). In a real sense, alcohol and drug abuse increases depression and anxiety and does not solve the situation. Arguably, those already in tobacco and opioid abuse show the most significant potential for worst outcomes when they get infected with the virus (Haddad et al., 2021). The reason behind this is that the above addiction is destroying the respiratory system, thereby weakening its immune system and causing chronic conditions, such as lung diseases and heart complications, that increase the risk of severe complications from COVID-19 (Haddad et al., 2021). These justified reasons make it necessary to learn self-care strategies for coping purposes other than relying on drugs to keep stress away.

Relevance of Understanding the Psychological Effects of the Pandemic

Understanding the psychological effect of the corona pandemic is of great importance to an individual and the general society (Pfefferbaum & North, 2020). Without this understanding, humanity cannot deal with the pandemic and its associated effects. Therefore, through analysis of the psychological effects of the pandemic, it will be possible to determine what approaches would work where and when to apply them.

An infection cannot be managed or contained if its roots and effects are unknown. People and concerned agencies can address the effects without any difficulty. The most special relevance is that it will be easy to understand the management schemes of the virus with the knowledge acquired.

Increasing awareness of the pandemic is relatively critical, as it allows them to have a wide range of points on the management of the pandemic (Kang et al., 2020). Awareness campaigns on the effects of the pandemic on the people at large are crucial to disease management and containment.

Information and convictions are fundamental in ailment portrayals (or composition), which drive conduct.

With the correct data on the disease, management approaches are efficiently conducted. The data must be accurate and reflective of what is there.

Disease assumptions built from information about the infection, how it is acquired, and what limits transmission might secure against possibly destructive convictions are dependent on false impressions and questions of trustworthy sources ((Pfefferbaum & North, 2020). For instance, it is understood that a quick transmission pace could overpower the medical services framework and thus limit care for everybody. Those with COVID-19 may keep people from accepting those individual decisions and should adhere to defensive practices. In this way, information-based sickness portrayals might fill in as a complete route device for settling on successful well-being-related choices during the pandemic.

Origin of Corona Virus, Risk Factors, Causes, and Effects

The world is slowly managing the COVID-19 pandemic, whose impact has amounted to immeasurable quantities. To dismiss the mythical arguments regarding the cause of the disease the World Health Organization (2020a) argued that the "virus that causes COVID-19 is in a family of viruses called Coronaviridae," and antibiotics are inadequate to control or prevent measures. As many people purported before, the disease is caused by a virus and not a bacterium. When someone is infected with severe acute respiratory syndrome coronavirus 2 (SARSCoV-2) virus strain, they become a victim of coronavirus. Numerous well-being specialists accept that the new strain of COVID probably started in bats or pangolins (Andersen et al., 2020). The primary transmission to people was in Wuhan, China. From that point forward, the infection has, for the most part, spread through individual-to-individual contact.

Around 80% of individuals with COVID-19 recuperate without expert treatment (Prescott & Girard, 2020). These individuals might encounter gentle, influenza-like manifestations. Be that as it may, one out of six individuals might experience extreme manifestations—for example, inconvenience relaxing (Anand et al., 2021). The new COVID has spread quickly in many areas of the planet. On March 11, 2020, the World Health Organization (WHO) pronounced COVID-19 a pandemic. A pandemic happens when a sickness that individuals are not resistant to spreads across huge districts.

SARS-CoV-2 spreads through close networks from one individual to another (Andersen et al., 2020). When individuals with COVID-19 inhale or hack, they dispel tiny drops containing the infection (Andersen et al., 2020). These droplets can enter the mouth or nose of somebody without the condition, making them infected. The most widely recognized way this disease spreads is through close contact with somebody who has the contamination (Lotfi et al., 2020). Close contact is inside around 6 feet. The sickness is most infectious when an individual's manifestations are at their pinnacle. Arguably, it is feasible for somebody without indications to spread the infection (Lotfi et al., 2020). A review by Singhal (2020) proposed that 10% of contaminations are from individuals showing no manifestations. Drops containing the infection can likewise arrive on adjacent surfaces. Others can get the disease by contacting these surfaces or articles. Contamination is reasonable if the individual constantly touches their nose, eyes, or mouth. Note that COVID-19 is new, and research is yet progressing. There may likewise be alternate ways that the new COVID can spread.

Most occasions of COVID-19 are not authentic. In any case, it can cause indications that become genuine, inciting passing from time to time. The scene of COVID-19 is startling. This makes it difficult to evaluate how much time the ailment becomes genuine or the speed of mortality. Singhal (2020) suggested that out of 1,099 people with assorted cases in China, around 16% became outrageous. Another report assesses that about 3.6% of the avowed cases in China died (Lotfi et al., 2020). These figures are likely going to change as the situation progresses. In any case, they suggest that COVID-19 is riskier than influenza. For example, infrequent influenza ordinarily prompts passing in under 0.1% of cases (Lotfi et al., 2020). When testing becomes less complicated and more expansive, prosperity experts will comprehend the number of outrageous cases and deaths. SARS is another kind of COVID. It transformed into an overall pandemic in 2002–2003. Around 9.6% of SARS cases provoked passing (Lotfi et al., 2020). COVID-19 is more irresistible and is currently the justification behind extra deaths throughout the planet.

Symptoms and Risk Factors

Very little was known about the novel COVID at the pandemic's beginning since specialists had not experienced it. Since clinical specialists have had the opportunity to concentrate on the effect of infection on patients throughout the planet, it is essential to acquire a superior comprehension of how it acts and what it means for various populations (Jeon et al., 2020). A developing assemblage of exploration shows that fundamental medical issues raise the danger of becoming genuinely sick or passing on from COVID-19 (Jeon et al., 2020). As indicated by one review, a few conditions might twofold or even triple the danger of death for COVID-19 patients. Understanding who is at expanded risk of genuine confusion helps specialists properly consider the virus.

Being infected with the virus depends on one's proximity to the infected individual (Lotfi et al., 2020). Being closer to a person infected with the virus increases a person's chances of acquiring the virus. Also, people at high stakes have constant contact with the infected, like healthcare providers and the affected families (Lotfi et al., 2020). Living in large and congested cities and urban centers is equally an increasing point of infection (Haseltine, 2021). Arguably, there is a crop of individuals with underlying healthcare conditions, such as chronic infections (Jeon et al., 2020). They are mostly the aged (Haseltine, 2021). When exposed to such viruses, they become sick and easily succumb to the infections. Notably, the underlying conditions in this sense include cancer, pregnancy, type 2 diabetes, obesity, smoking, heart conditions (including heart failure), coronary artery disease or cardiomyopathies, and poor immunity due to organ transplants (Jeon et al., 2020).

As identified above, the risk of getting the virus, apart from having underlying medical conditions, is direct contact with the infected person (Jeon et al., 2020). Infected individuals will easily transmit the virus from one person to the other through body contact, which must be considered when attempting to reduce the infection levels (Haseltine, 2021). As per the CDC in Jean et al. (2020), hospitalizations have been multiple times higher and deaths multiple times higher among those with primary revealed conditions contrasted and the individuals who did not have fundamental ailments. A detailed review found that among over 900,000 individuals hospitalized for COVID-19 in the U.S. through November 2020, almost 63.5% were inferable from four conditions: weight, hypertension, diabetes, and cardiovascular breakdown (Vahey et al., 2021). Notably, these hospitalizations probably would not have happened if these conditions had not been available. The hazard expanded with age (Haseltine, 2021). Moreover, inside any age bunch, COVID-19 hospitalizations inferable from every ailment were higher in Black individuals and White grown-ups.

All that is known highlights the significance of keeping up with well-being-advancing practices, even as limitations ease, particularly when one has a hidden ailment (Jean et al., 2020). To diminish the danger of becoming ill, one should be immunized, wear a mask, keep an actual separation of at least six feet when around others, and maintain hygiene (Vahey et al., 2021). All these highlight the significance of keeping up with well-being-advancing practices, even as limitations ease, particularly if one has a hidden ailment. In August 2021, the FDA approved a third portion of mRNA immunization (Pfizer/BioNTech or Moderna) for specific immunocompromised individuals. An individual with a debilitated insusceptible reaction is bound to have advanced contamination and is at an expanded danger of more severe illness if they become tainted (Haseltine, 2021). A third immunization portion might upgrade specific immunocompromised individuals' resistance reactions (Vahey et al., 2021). Individuals should converse with the primary care physician if they figure they might fit the bill for a third COVID immunization portion.

The effects of COVID-19 on different people vary. Most infected people who develop mild to moderate illnesses do not require hospitalization to recover. The most common symptoms, according to Haseltine (2021) and Vahey et al. (2021), include:

- fever
- cough
- tiredness
- loss of taste or smell

Also, the following symptoms are less likely to be observed in patients:

- sore throat
- headache
- aches and pains
- diarrhea
- a rash on the skin or discoloration of fingers or toes
- red or irritated eyes

The severe symptoms include difficulty breathing, chest pain, and speech and mobility loss (Andersen et al., 2020). When one experiences these symptoms, seeking medical attention for critical conditions that become hard to control is essential.

Hazard of, and versatility to, poor psychological well-being changes across populace gatherings, eminently by age, sex, work status, monetary circumstance, and socioeconomic status (Andersen et al., 2020). Individuals with less secure businesses, lower instructive status, and lower pay encounter higher mental trouble during the COVID-19 emergency (Andersen et al., 2020). Youngsters, those living alone, those with a lower financial situation, and jobless individuals had higher paces of mental pain (Andersen et al., 2020). Then again, individuals who had the option to keep working during control, including those who worked distantly, were more averse to reporting sorrow and tension, essentially during the previous phases of the emergency.

Lower financial status has been related to an expanded danger of poor psychological wellness (Bialowolski et al., 2021). During the COVID-19 pandemic, this and impeded gatherings predicted individuals encountering higher paces of mental misery. In the United Kingdom, a tension score was estimated persistently for 20 weeks from the beginning of the principal lockdown in March 2020; higher uneasiness scores were reliably detailed among individuals with lower training or lower pay (Weida et al., 2020). One more report from the United Kingdom found that individuals whose pay dropped became reliant on various types of state support (Andersen et al., 2020). For example, all-inclusive credit or independent work awards—or individuals encountering long-term monetary troubles—experienced more noticeable decreases in their psychological well-being than the normal populace during the main lockdown in April 2020 (Andersen et al., 2020).

Nonetheless, patterns in changing psychological wellness status are unreliable across all populaces. It has been challenging to survey whether the emotional well-being of individuals with lower SES has deteriorated quicker or more contrasted with populace midpoints. For instance, in the United States, a study in April 2020 discovered that people with higher instructive fulfillment and family pay detailed, sharp decreases in life fulfillment and more excellent expansions in burdensome side effects than individuals with lower SES, contrasted with overview results in 2019 (Weida et al., 2020).

On 30 December 2019, the Program for Monitoring Emerging Diseases advised the world about pneumonia of obscure reason in Wuhan, China (She et al., 2020). From that point forward, researchers have gained tremendous headway in understanding the causative specialties in the extremely intense respiratory disorder COVID-2 (SARS-CoV-2) and its transmission, pathogenesis, and relief by antibodies, therapeutics, and non-drug intercessions. However, more examination is expected to decide the beginning of the pandemic (Maxmen & Mallapaty, 2021). Hypotheses of unplanned delivery from a lab and zoonotic overflow both stay practical. Knowing how COVID-19 arose is essential for educating worldwide on techniques to alleviate the danger of future episodes.

With the origin traced to Wuhan, China, it then means that the virus was carried to other parts of the world. The Wuhan Institute of Virology is in a region where the studied viruses are endemic (Maxmen & Mallapaty, 2021). Much research is conducted at the Wuhan Institute of Virology to develop the best solution that can be used to address the associated effects of the virus. Most researchers say SARS-CoV-2 presumably has a characteristic beginning and was sent from a creature to people. Nonetheless, a lab spill has not been precluded. Many require a more profound examination concerning the speculation that the infection arose from the Wuhan Institute of Virology (WIV), situated in the Chinese city where the primary COVID-19 cases were accounted. On 26 May, U.S. President Joe Biden entrusted the U.S.

Intelligence Community to join endeavors to discover SARS-CoV-2's starting points, whatever they may be, and report back in 90 days (Maxmen & Mallapaty, 2021). Australia, the European Union, and Japan have likewise required an intense examination concerning SARS-CoV-2's beginnings in China (She et al., 2020). The WHO presently cannot uncover the following period of its review.

Nevertheless, China has asked that the test looks at different nations. Such hesitance around how China has retained data before fueled doubts about a leak from the labs. For example, Chinese government authorities smothered critical general well-being information toward the beginning of the COVID-19 pandemic and during the 2002–2004 serious, intense respiratory condition (SARS) scourge, per undeniable level reports.

Researchers need more proof regarding the beginnings of SARS-CoV-2 to preclude the lab-spill speculation or to demonstrate the other option—that the infection had a characteristic beginning (Maxmen & Mallapaty, 2021). Numerous irresistible infection specialists concur that the situation is that the disease advances typically and spreads from a bat either straightforwardly to an individual or through a halfway creature. Most irresistible sicknesses start with an overflow from nature, as was seen with HIV, flu plagues, Ebola flare-ups, the COVID that caused the SARS scourge beginning in 2002, and the Middle East respiratory condition (MERS) episode starting in 2012.

In principle, COVID-19 might have come from a lab in several ways. Scientists may have gathered SARS-CoV-2 from a creature and kept up with it in their lab to study, or they may have made it by designing COVID genomes (Maxmen & Mallapaty, 2021). In these situations, an individual in the lab may have been accidentally or intentionally contaminated by the infection and afterward spread it to others, starting the pandemic (Fenn, 2020). There is no evident proof to back these situations; however, they are not unimaginable. Individuals have made various contentions for a lab beginning for SARS-CoV-2 that are, at present, a guess. One holds that it is dubious that, 12 and a half months into the pandemic, SARS-CoV-2's nearest relative has not been found in a creature. Another proposes it is no small occurrence that COVID-19 was first recognized in Wuhan, where a top lab considering COVIDs, the WIV, was found (Fenn, 2020). Some lab spill advocates fight that the infection contains strange elements and hereditary successions, flagging that people designed it. Also, some say that SARS-CoV-2 spreads among individuals so promptly that it has been made with that expectation

(Maxmen & Mallapaty, 2021). Another contention recommends that SARS-CoV-2 may have gotten from COVID found in an unused mine where WIV scientists gathered examples from bats somewhere in the range between 2012 and 2015.

Discussion Activity

So, what do irresistible infection scientists and developmental researchers say about these contentions? This section will discuss some of the theories, myths, and foundational knowledge experts have heard about the origin and spread of the virus.

Relationship Between the Virus and Mental Health

Since the infections from COVID-19 were declared a pandemic problem, many people have succumbed, and many more have acquired deep mental health issues. The pandemic's conditions force people to live and make life more stressful and unworthy of living (OECD, 2021). The pandemic and the current resulting economic recession have negative consequences on the general mental health of many people (Shukla, 2020).

People with underlying mental disorders and drug abuse sprees have been pushed further into their depression as the conditions brought about by the virus make it hard to cope with their lifestyle (OECD, 2021). More than 40% of adults in America have indicated anxiety and depressive disorder symptoms (Shukla, 2020). The trend continues to rise as time, and the effects of the pandemic continue to be felt. Among the conditions experienced by many people include difficulty sleeping, and eating, an increase in alcohol consumption or substance use, and worsening chronic diseases because of the stressful episodes of the corona pandemic (OECD, 2021). As the pandemic increases, ongoing and needed public health measures are needed to avoid exposing several people to experiencing situations connected to poor mental health outcomes, such as isolation and job loss (OECD, 2021). Separate groups within the demographics are affected differently, and the mental health issues associated with the pandemic are also different.

The above paragraph indicates that the virus's effect indirectly impacts people's mental health. It results in conditions that worsen people's normal operations (Shukla, 2020). When one is denied basic human requirements based on survival requirements, some take a long time to adjust to the new normal. The effect is depression and other associated mental health disorders (OECD, 2021). For instance, many people are forced to avoid gatherings because of the pandemic since they could easily acquire the infection. Avoiding crowds implies no parties, no church meetings, and an isolated lifestyle. Worst, school-age children have no classes, and onsite learning is disrupted. People are constantly forced to be isolated, and there is a continued call to remain isolated whenever one thinks they might have acquired the virus. All these bring about mental health problems.

Whether one stresses over oneself, family, companions, or associates, there is no moving away from the dread of becoming contaminated with the COVID-19 infection (OECD, 2021). Progressing dread is a side effect of tension, which influences an individual's craving, messes rest up, and impacts their capacity to adapt to their routine (Singhal, 2020). Dread also prompts different wellsprings of stress, like filling practices, such as accumulating food (Shukla, 2020). Regardless of whether one needs to hoard, one will be disappointed in not tracking down the fundamental things one needs at the store (Singhal, 2020). Arguably, remaining at home should calm a portion of someone's feelings of trepidation about becoming ill; however, disconnection accompanies its aftermath (Shukla, 2020).

Social distancing isolates an individual from the help and friendship of loved ones, which is amazingly troublesome and discouraging. Regardless of whether one is at home with family, the truth of social confinement can, in any case, trigger melancholy, bitterness, and uneasiness (Singhal, 2020). Likewise, people may find spending the entire day with their family unpleasant and difficult, regardless of how much they love their family.

Discursively, the pandemic has led to a loss of jobs as many people are laid off (Varshney et al., 2020). It brings about financial worry. Even with a robust monetary base, the uncertainty of when one gets back to work or whether the job will be available are overwhelming stressors contributing to mental health problems (Varshney et al., 2020). Also, there is grief due to losing loved ones (Singhal, 2020). Several people have experienced or will soon suffer the deep grief caused by the sudden death of friends and family. However, grief is a reaction to any loss, which means the COVID-19 pandemic may also cause grief over losing one's job or health (Singhal, 2020). For quite a long time, the commonness of emotional wellness conditions has been extensively unaltered; this pattern changed in 2020 with the episode of the COVID-19 pandemic.

From March 2020 onwards, the pervasiveness of uneasiness and sadness expanded. For instance, in Belgium, France, Italy, Mexico, New Zealand, the United Kingdom, and the United States, the pervasiveness of uneasiness in mid-2020 was twofold, or over two times the level seen in earlier years. In Australia, Belgium, Canada, France, the Czech Republic, Mexico, Sweden, the United Kingdom, and the United States, the predominance of wretchedness in mid-2020 was twofold or over two times what was seen in earlier years (OECD, 2021). When contrasted with the outcomes before the pandemic, the information reliably shows that paces of uneasiness and sadness expanded in 2020 compared to earlier years. Across three nations, distinctive review instruments have been utilized to quantify the predominance of tension, sorrow, and the degrees of mental misery (OECD, 2021). The commonness rates and the progressions in these rates are not straightforwardly tantamount across nations, and some reviews embraced during 2020 do not cover broadly delegated tests (Jean et al., 2020). Regardless, the predominance of tension and melancholy has unmistakably expanded altogether (Singhal, 2020). During 2020, the danger factors for poor emotional well-being—monetary instability, joblessness, and dread— developed, while defensive components—social association, work, instructive commitment, admittance to actual exercise, day-by-day schedule, and admittance to well-being administrations—diminished (OECD, 2021). It is evident that the effects of the pandemic are not negligible, and they will for a long time live with people. Therefore, it is essential to understand coping strategies while minimizing the potential impact on the affected psychological well-being.

Chapter 2: Psychological Effects on Various Professionals

The psychological impacts of COVID-19 do not only manifest in each population, but they affect people in various capacities and various ways. While the effects may be similar, the way they manifest themselves and the long-run impact on multiple people is different. In this section, the effect of the pandemic on various population sects and professions will be analyzed. The major population sects that will be examined and how the pandemic affects them include children and students, nurses, healthcare workers, and the elderly. The other population sector that will not be analyzed will be assumed to be undergoing the generalized psychological effects of the pandemic. However, it is noteworthy that the pandemic impacts every population sector and professional field differently. This implies that the assumed psychological impact of the pandemic will follow from the generalized effect on the given area.

Psychological Effects on Students

The learning institutions are points where students from all races meet and learn about various life experiences. Students constantly undergo much mental pressure whenever they are shut down due to pandemics or natural calamities, and some of them end up messed up since they cannot just find a point to meet and share their life experiences (Browning et al., 2020). Therefore, many institutions have been closed since the pandemic struck, and this effect is paramount. Closure of the learning institutions is majorly done to address the spread of the virus.

In general, university students and most high school students are considered vulnerable since most suffer from anxiety, depression, substance abuse, and no uniform eating habits. Therefore, whenever something affects their educational schedule, such as the pandemic, most of them find their mental stability issues amplified, and it becomes hard for them to cope (Browning et al., 2020). Some have long-term effects and post-traumatic stress disorders, as they hardly adapt to the new normal. The impact of the pandemic is highly pronounced among college and university students since their academic success, social college life, future careers, and other factors are uncertain (Chen et al., 2020). Notably, even before COVID-19 was declared a pandemic, students globally indicated increased cases of anxiety, depressive moods, lack of self-esteem, psychosomatic problems, substance abuse, and suicidal intentions (Browning et al., 2020). It implies that when dealing with the students, it is essential to input more resources to help identify and solve the pandemic's physical and mental consequences.

Students have indicated an increased lack of motivation in their studies, as they are battling several factors that make them lack priority. According to Browning et al. (2021), students from various universities and colleges across the world indicated poor motivation to undertake their studies due to the unpredictability of their future careers. As observed, many people go jobless in the fields and are still studying. This reduced their morale to do more in their academics, as they believed the same fate would befall them (Chen et al., 2020). Many students found it necessary to divert attention to important issues, such as their health, where the motivation was pushed. It is more essential to be healthy and free from infection than acquire academic excellence and get sick.

Arguably, the pandemic has resulted in isolation that directly consequences loneliness and anxiety (Jean et al., 2020). Isolation is one of the recommended ways of managing the spread of the disease from one person to the other. However, whenever students are isolated, especially the international ones or those suspected of being exposed to the disease, they become psychologically disturbed (Chen et al., 2020). They lack others to talk to and people to share some experiences with one-on-one. An isolated individual may think they are unworthy and start self-stigmatization as an island. Stigma and discrimination can also worsen someone's mental health problems and delay or stop them from getting help. Social isolation in poor conditions, unemployment risks, and poverty are linked to mental health. So, stigma and discrimination can trap people in a cycle of illness.

Discursively, stress and anxiety are all attributed to the pandemic effect on students. Stressful episodes result in fear, as the students will always want to know what the future holds. Stress results from the unpredicted future of academic life due to the pandemic and the fact that it is hard to know when regular studies will resume.

Intervention Recommendations

With the considerable level of understudies appointed to the high mental effect profile, colleges would be very well served to address the emotional wellness needs of the entire student population since all may be experiencing the difficulties. Select projects that have advanced psychological wellness, for example, are virtual gathering activities and contemplation or care meetings, responsibility pals and exercise difficulties, medication, and advising visits. These gatherings might help bring down uneasiness and diminish the feeling of segregation revealed by the understudies in this review. Advanced intercessions for understudies with clinical degrees of nervousness or discouragement can include computerized and mixed remedial mediations to mitigate the potential for self-damage or self-destruction. Calls and instant messages arrive at those with less automatic assets, self-destruction hazard evaluations, chatlines and gatherings, and different advances to screen hazards either inactively or effectively. It is a working plan to apply the six-step process of solving mental illness among students, as indicated in Chen et al.'s (2020) study. These steps include conveying positive pandemic-related data, decreasing lousy conduct, finding out about pressure on the board methods, upgrading family connections, expanding in specific behavior, and changing scholarly assumptions.

Arguably, with the probability of progressing mental trouble from COVID-19, colleges may likewise consider assisting understudies with keeping up with solid attitudes instead of staying away from stress (Bavel et al., 2020). On the side of this suggestion are late discoveries that intellectual and conduct evasion, like staying away from circumstances where openness is conceivable and one may encounter troublesome considerations about the pandemic, was the steadiest indicator of expanded uneasiness and burdensome side effects of the pandemic. Intellectual reappraisal of unpleasant circumstances can adjust their adverse consequences (Crum et al., 2020). Versatile outlooks can likewise assist with revamping needs to foster further connections and more considerable enthusiasm for life, just to assist understudies with acclimating to better approaches to learning. Since some of the understudies in this review were inclined to be less persuaded, functional, and ready to center, changing to a versatile outlook might assist understudies with persisting in their schooling now and sometime down the road. At last, mentality reappraisals can build prosperity, decline negative well-being manifestations, and lift physiological working under intense pressure when a relative becomes contaminated. The pandemic necessitates quick changes in approaches and methodology that influence understudies.

Colleges can additionally foster stages that work with safe understudy social cooperation. Numerous understudies search for friendly cooperation during their college experience. Nonetheless, as the discoveries of this review uncovered, understudies' chances for mingling altogether diminished in the beginning phases of COVID-19 (Crum et al., 2020). Missing going out and significant achievement occasions were common reactions from understudy members. Different examinations tracked that down to keep up with understudies' emotional well-being during the principal wave of the COVID-19 pandemic. They imparted online with close relatives or flat mates day by day. With understudies, physical separation does not and ought not to need isolation. Zoom and offbeat Facebook bunch online associations can cultivate holding and spanning social association, stretching out past web-based media posts and email list servers (Jones et al., 2020). Typical settings where individuals gather, like spots of love, rec centers, cafeterias, yoga studios, and classes, can be duplicated on the web or even held outside in a calm climate on a timetable, like what was set up before the pandemic. As of late, effective mediation incorporate the internet sharing of plans, books, and digital broadcasts, such as virtual films, games, random data, or party time evenings (Jones et al., 2020). Offering help to understudy associations to arrange these virtual social exercises could speed up the accessibility of these assets.

Effects on Nurses and Healthcare Workers (HCWs)

Healthcare professionals are at the center of managing the pandemic, and if there are any people highly affected by the pandemic, it is the healthcare providers (Asnakew et al., 2021). Whenever healthcare workers and nurses undergo psychological discomfort because of the pandemic, the quality of care will decrease, and the healthcare system will be characterized as poor (Asnakew et al., 2021). Therefore, it is essential to analyze how the pandemic affects healthcare workers mentally and what should be done to help solve the problem. In this sub-section, the analysis will entail identifying potential causes of mental illness during the care process and what can be done to reduce the impact of the problems.

Fear of Contracting the Disease

Healthcare workers and nurses who handle the victims of COVID-19 are highly exposed to the disease (Asnakew et al., 2021). The probability of acquiring the virus is high, bringing about some fear in healthcare providers, especially in conditions where Personal Protective Equipment (PPEs) is not available or adequate (Asnakew et al., 2021). People exhibit different feelings of fear, worry, anxiety, depression, and stress due to the continuous updates on the spread and infections of COVID-19 (Kang et al., 2020). Healthcare workers are at high risk of disease—more than any other population segment, as they directly handle the patients having the virus (Asnakew et al., 2021). It makes them fear contagion, be concerned with family health, experience interpersonal isolation, desire trust in and support from their organization, and encounter stigma (Asnakew et al., 2021). This is the overwhelming pressure on psychology, which will cause mental-related illnesses such as anxiety, depression, fear, and stress (Kang et al., 2020). Therefore, creating an awareness campaign that will aid in protecting the frontline healthcare workers from viral exposure and its associated consequences is essential. The new standard introduced by the pandemic had both physical and mental health impacts on healthcare workers during the COVID-19 pandemic.

Inadequate and Ineffective PPE

When personal protective equipment is not enough, and its effectiveness is questionable, it becomes a mind-boggling and worrying experience for healthcare providers (Kang et al., 2020). Ineffective PPEs will increase an individual's exposure to viral infection (Kang et al., 2020). Therefore, it is essential to ensure that healthcare workers are not exposed to the virus by providing adequate and effective PPEs.

Other Aspects

In a study conducted by Asnakew et al. (2021) relating to discouragement, respondents who had clinical sickness had higher chances of misery than those who had no clinical disease, which was upheld by past studies. Respondents who had a background marked by psychological maladjustment were almost certainly influenced by sorrow as contrasted with their partners. This study's results were like those found in a similar survey conducted in Nepal (Khanal et al., 2020). The nurses and other healthcare workers who had contact with the affirmed COVID-19 virus were more influenced by depression, anxiety, and stress than their partners (Zheng et al., 2021). The past investigations upheld this led China to audit the effect of COVID-19 on the HCWs quickly

(Zheng et al., 2021).

As of now, evidence from past investigations tracked down the beneficial outcome of social help on the emotional well-being of the HCWs. Additionally, the discoveries were reliable in past studies. It was discovered that straightforwardly reaching the suspected or affirmed COVID patients was the significant danger factor that caused mental misery and uneasiness among the HCWs (Zheng et al., 2021). Moreover, helpless social help has a fundamental connection with sorrow, anxiety, and stress (Spoorthy et al., 2020). Besides, it is notable that good social support given by accomplices or life partners and companions addresses a defensive factor for mental well-being (Spoorthy et al., 2020). The absence of help to make up for actual inadequacy, passionate use, and somebody to converse with about the pandemic or go to for exhortation could build the danger of emotional wellness issues (Kang et al., 2020). Of all these factors, women are at high risk of being stressed, depressed, and anxious from the pandemic's effects. A study conducted in India by Wilson et al. (2020) found that women are at high risk based on how they perceive the infection relative to their social roles.

Approach to Addressing the Issues

Healthcare workers and nurses are influenced by the undeniable degree of tension, stress, and discouragement during the COVID-19 pandemic. The COVID-19 pandemic has disturbed the emotional well-being of the populace, especially the medical care laborers, and the interest in psychological well-being administration is expanding. The studies indicate the requirement for top caliber, hypothesis, and proof-based mediation for the prosperity of medical care laborers during this COVID-19 pandemic. This concentration likewise portrayed the need for emotionally supportive networks and adapting techniques that might help to decrease pressure, uneasiness, and melancholy among medical care laborers. These methodologies might target female HCWs with a clinical ailment, a history of psychological maladjustment, contact with affirmed COVID-19 patients who need social help, and families with the persistent disease during their everyday practice. In addition, dealing with the HCWs is essential to reinforce their emotional wellness, which adds to the nature of clinical benefits as a forefront warrior during this pandemic.

Effects on the Elderly

Older adults are highly at risk of mental illness and exhibit depressive symptoms (Mukhtar, 2020). However, the pandemic has worsened since it creates highly conducive conditions for depression, anxiety, and stress in this vulnerable population (Mukhtar, 2020). Unfortunately, managing the pandemic requires self-isolation, no gatherings, and reduced contact between people in different settings (Kang et al., 2020). These factors increase loneliness among the old, making them feel neglected. In systems and cultures where group meetings, group eating, and joint sleep-over are encouraged to show care for the elderly, the chances of the elderly being forced to adapt to a different lifestyle increase stress and depressive disorders. Arguably, social isolation, social disconnectedness, and loneliness among the elderly increase depression and anxiety (Mukhtar,

2020). Notably, since most elderly have chronic health conditions, most of them succumb to the infection, and whenever this news reaches the living, they tend to lose hope and become afraid, knowing they are next in line to die.

Solution

The dependence on online media could go about as an apparatus to forestall sadness, weariness, and dreariness in youthful gatherings; however, the need for social help, energy, and day-by-day work remains neglected for the more established age bunch. Online innovations and computerized sources are currently outfitted to offer virtual-advanced social help organizations and an apparent feeling of having a place. The differences in admittance to the proficiency of these cutting-edge details are lost at a large portion of the old gathering population. The mass quarantine and transport restriction have choked more established grown-ups' exercises: standard walk-and-talk in the recreation center; association meetings; intentional assistance and social consideration; congregational get-togethers; restricted contact with plants and creatures; and getting the recommended sustenance, prescription, and treatment. Accordingly, further disturbing difficulties arose in the wake of COVID-19 for the emotional wellness of more established grown-ups locally. Lacking and deficient consideration has been paid to the emotional health of the more established age bunch as far as ideal and quality mental emergency intercession.

Activity-based mental preventives and general well-being methodologies could develop social associations and advance solid associations with oneself and others. Intellectual abilities and encouraging social groups of people could assist more seasoned grown-ups with cultivating significant association and feelings of belongingness during the disconnection period. Intelligent conduct and friendly, positive, and brief treatments conveyed on the web or in person could upgrade mental prosperity and work on close alliances while at the same time decreasing apparent forlornness. Social seclusion and social detachment are an archived bidirectional and complex connection between psychological well-being issues and social disjunction. It represents genuine general well-being worries among more seasoned grown-ups, mainly due to the psychosocial reasons and physiological medical issues of different types, like emotional wellness, cardiovascular, immune system, neuro-cognitive, neurobiological, and other dangerous medical issues. The concerned agencies should take substantial guidelines for senior individuals socially confined at home or isolated at medical care offices (medical clinic, facility, disconnection unit, childcare, public venue, and spots of love). They need to have endorsed diet and prescriptions and have conveyed the importance of social in-touch to relieve their physical and emotional wellness consequences. However, adherence to social segregation methodologies could be debilitated with time. Such much-planned supported preventive measures would proficiently forestall the bothered bleakness of COVID-19, identified with psychological well-being issues in more established grown-ups.

The conditions will arise at some point, and an individual's readiness to tackle them determines if they will come out strong. The pandemic's psychological effects require one to be ready for self-care to avoid or manage the conditions. The following self-care strategies are essential and should be embraced.

Mitigating the Psychological Impact of COVID-19 on Healthcare Workers

The psychological effects of the pandemic will not fade, even after the control of the pandemic, since they have deeply afflicted pain in people. The mental effects on workers have adverse associations (Kang et al., 2020).

The outrageous tensions experienced by medical care laborers during a pandemic might build their danger of burnout, which has unfriendly results for individual prosperity, patient consideration, and the medical services framework. During the episode, feelings experienced by medical services laborers were related to renunciations and helpless work execution.

HCWs with a greater danger of openness to the infection experience a more remarkable mental effect than those with less transparency (Pfefferbaum & North, 2020). In China, bleeding-edge medical care laborers that focused straightforwardly on patients with COVID-19 experienced pressure, uneasiness, and a sleeping disorder and displayed more elevated levels of emotionally severe well-being manifestations than those in optional jobs (Shanafelt et al., 2020). Then again, different investigations have shown a higher prevalence of mental misery among non-forefront staff, perhaps because these specialists have less admittance to data and mental help (Pfefferbaum & North, 2020). These features that help systems are required for all medical services laborers, regardless of their work job or level of infection openness.

Mentally supporting the HCWs and any individual impacted by the pandemic should zero in on hierarchical just as individual qualities, with a more extensive objective of keeping a traditional culture of flexibility (Douglas et al., 2020). Earlier pandemics have exhibited that the setting of the association affects mental results in the labor force. It is grounded that social standards inside an association, such as initiative styles and executive correspondence examples, are critical variables in specialist stress (Shanafelt et al., 2020). In pandemic circumstances, clear orders and careful steps decrease the probability of enthusiastic misery as friends support. Social help outside of the work environment may likewise cradle pressure. However, medical care laborers regularly disregard associations with their loved ones because of big jobs or worry about contaminating others because of their word-related openness to the infection (Shanafelt et al., 2020). Support of social contact is progressively tricky regarding removing social prerequisites. Narratively, there are reports of medical care laborers encountering social shame and maltreatment because of public apprehensions about getting the infection from those with the most apparent openness.

During the pandemic, five common phrases were applied by the people undergoing psychological issues: *hear me, protect me, prepare me, support me,* and *care for me* (Shanafelt et al., 2020). These phrases are vital in determining what is required to mitigate the effect of the pandemic. Taking these phrases for granted whenever one hears them is the initial point of making the person lose hope and become more desperate (Shanafelt et al., 2020). For the medical services labor force to perform to their maximum capacity throughout a drawn-out time, medical services bosses should give early psychosocial backing to all workers that tend to these solicitations that are centered around the making of a mentally protected climate, solid administration, clear hierarchical procedures for staff prosperity, reliable correspondence, and tremendous group support (Shanafelt et al., 2020). Such a climate will cultivate individual versatility, approval, self-empathy, and self-care. Building a culture of hierarchical versatility might assist with decreasing the probability or seriousness of mental indications requiring treatment once the immediate danger of COVID-19 dies down.

Additionally, Blake et al. (2020) proposed a digital learning package that will offer online lessons to people and help them with great insights on how they can apply to manage the pandemic effect. This package has several ideas on how one may address the associated psychological effects of the pandemic. While the box called "Psychological well-being in Healthcare Workers:

Mitigating the Impacts of COVID-19" has its origin in the U.K. It can be applied in vast areas and regions. Also, despite being formulated for HCWs, it can have broader applicability to people going through various psychological effects of the pandemic.

Mitigating the Psychological Impact of COVID-19 on Children

Traumatic events have momentary psychological well-being consequences, yet the drawn-out impacts can be similarly crippling. Mounting proof has recommended that the effects of horrendous events wait long after the actual occasion and can prompt unfavorable well-being results, including heftiness, hypertension, and cardiovascular infection (Shanafelt et al., 2020). Accordingly, we should focus on understanding the psychological wellness impacts of the pandemic on kids.

As society gradually gets back to its post-pandemic form of typical, we should recall that the consequences for psychological well-being wait in our kids and may have their drawn-out capacities ultimately work as the kids become grown-ups. We should be proactive in ensuring the emotional wellness of our youngsters (Shanafelt et al., 2020). Put resources into and vigorously advance dynamic well-being administrations for youngsters, considering networks of shading—explicitly Native American, Black, and Hispanic people groups—as these kids experienced the brunt of the pandemic. They are additionally more averse to approach care. The National Institute of Mental Health (NIMH) should focus on this. Grown-ups, not youngsters, are currently leading NIMH's studies.

The approach should be to develop a public associate of small kids to concentrate on the psychological well-being impacts of the pandemic over the long haul, guaranteeing that youngsters from all networks are very much addressed. Also, the concerned agencies should support the capacity of schools and youth sports associations to recognize battling kids and give assets to assist with increasing mental pressure. Consequently, it is essential to proactively tend to the emotional well-being issues among youngsters, which can go far toward working on the soundness of the up-and-coming age of pioneers. In any case, experts should act now, before it is past the point of no return.

Vaccination Against COVID-19

Immunizations permit people to meet up, side by side—in schools, networks, and places of love (Shanafelt et al., 2020). They are the best expectation of finishing the COVID-19 pandemic and returning to doing what everyone appreciates with their loved ones. Several vaccines have been created, but Moderna, Johnson & Johnson (Janssen), and Pfizer-BioNTech are the USA's three approved. They are all highly effective, provided one gets the correct dosage.

Myths and Truth About the Vaccines

Myth: The COVID-19 vaccine is dangerous because it was grown so rapidly.

Truth: The vaccines are demonstrated to be safe and effective. Although they were created in record time, they have gone through a similar thorough Food and Drug Administration process as different antibodies, satisfying all security guidelines (Mallapaty, 2021). No means were skipped. All things being equal, we can thank the extraordinary overall coordinated effort and speculation for the more limited period on the advancement of the antibodies. The clinical preliminaries and security audits took similar time measures as different antibodies.

Myth: The COVID-19 immunization will modify my DNA.

Truth: The primary antibodies contain courier RNA (mRNA), which teaches cells to make the "spike protein" found in the new COVID (Jean et al., 2020). When the safe framework perceives this protein, it constructs a safe reaction by making antibodies and showing the body defense against future contamination. The mRNA never enters the core of the cell, which is the place where DNA (hereditary material) is kept. The body disposes of the mRNA before it is completed by utilizing the guidelines.

Myth: The COVID-19 immunization incorporates a GPS beacon.

Truth: A video shared on many occasions on Facebook makes bogus cases about the results of needle creator Apiject Systems of America, which agrees with the public authority to give clinical-grade infusion gadgets for immunizations. The organization has a discretionary form of its item that holds a computer chip inside the needle name that assists suppliers with affirming an immunization portion's starting point. The actual chip is not infused into the individual getting the immunization.

Myth: The COVID-19 antibody has serious incidental effects like hypersensitive responses.

Truth: Some members in the antibody clinical preliminaries detailed aftereffects like those accomplished with different immunizations, including muscle agony, chills, and migraine. Also, albeit incredibly uncommon, individuals can have extremely hypersensitive responses to fixings utilized in an antibody. That is why specialists suggest that individuals with a background marked by robust, unfavorably susceptible responses, like hypersensitivity to the elements of the antibody, ought not to get the immunization.

Myth: COVID-19 immunization causes

Barrenness in ladies: Misinformation via Web-based media proposes that the antibody prepares the body to assault syncytin-1, a placenta protein that could prompt barrenness in ladies (Mallapaty, 2021). In all actuality, there is a corrosive amino grouping divided among the spike protein and a placental protein; notwithstanding, specialists say it is too short even to consider setting off an insusceptible reaction and consequently does not influence richness (Tartof et al., 2021).

Myth: I have effectively been determined to have COVID-19, so I do not have to get the antibody.

Truth: If one has effectively had COVID-19, there is proof that they can, in any case, profit from the antibody. The resistance somebody gains from having a disease, called normal invulnerability, shifts from one individual to another (Mallapaty, 2021). Some early proof proposes that normal invulnerability may not keep going extremely long. Specialists do not have the foggiest idea of how long a person is shielded from becoming ill again after recuperating from COVID-19 (Mallapaty, 2021).

Myth: Once I get the COVID-19 immunization, I, at this point, do not have to wear a mask.

Truth: Masking, handwashing, and physical removal stay vital out in the open until there are an adequate number of resistant individuals (Jean et al., 2021). Thoroughly inoculated individuals can meet with other completely immunized individuals without wearing covers. People must dive more deeply into the wholly inoculated life.

Myth: You can get COVID-19 from the vaccine.

Truth: You cannot get COVID-19 from the antibody since it does not contain the live infection

(Mallapaty, 2021).

Myth: Once I get the antibody, I will test positive for COVID-19.

Truth: Viral tests used to analyze COVID-19 check tests from the respiratory framework for the presence of the infection that causes COVID-19 (Mallapaty, 2021). The immunizations will not influence one's test outcome since there is no live infection in the antibodies. It is feasible to get tainted with the disease before the vaccine has had the opportunity to ensure one's body ultimately.

Myth: I am not in danger of serious confusion about COVID-19, so I need not bother with the antibody.

Truth: Regardless of one's danger, one can, in any case, get the disease and spread it to other people, so it is significant that they get inoculated (Mallapaty, 2021). It is not exclusively to secure one's family and the local area. When the antibody is broadly accessible, it has been suggested that however many qualified grown-ups as could be expected under the circumstances should get the immunization.

Myth: If I get the COVID-19 immunization, I will likely become wiped out from another ailment.

Truth: No proof getting the immunization uplifts one's danger of becoming debilitated from another contamination like influenza (Piltch-Loeb et al., 2021).

Myth: Certain blood classifications have fewer COVID-19 severe diseases, so getting an immunization is not required.

Truth: Research has displayed no excuse to accept that a specific blood classification will prompt the expanded seriousness of COVID-19 (Piltch-Loeb et al.,

2021). One can secure themselves, their family, and their local area by choosing to get immunized.

Effectiveness of Vaccine in Italy, India, and the USA

Antibody adequacy in forestalling SARS-CoV2 contaminations is 85.5% (95%CI: 75.9-91.3) in the to some-extent inoculated and 84.8% (95%CI: 73.2-91.4) i The wholly immunized (Piltch-Loeb et al., 2021). In forestalling, indicative contamination adequacy is 81.7% (95%CI: 62.7-91.0) n those who are inoculated to some extent and 87.1% (95%CI: 69.3-94.6) in the completely inoculated (Piltch-Loeb et al., 2021). The Covaxin vaccine is 81% effective (Mallapaty, 2021). The three vaccines' effectiveness is above 70% (Tartof et al., 2021). There were no COVID-19-related hospitalizations in those inoculated to some extent or wholly vaccinated versus hospitalization in the unvaccinated associates. Most results show effective and dependable outcomes against emergency clinic affirmations recently after being fully inoculated in an enormous, different population under proper immunization conditions, even despite boundless dispersal of the delta variation. These discoveries highlight the significance of further developing COVID-19 inoculation rates and remembering to deploy them where needed.

Viability against contaminations was high before long full immunization, both for delta and different variations of concern. However, it wound down over the review timeframe. Albeit winding down adequacy against emergency clinic confirmations was not seen in our review populace to date, this change should be cautiously monitored. After some time, there is a vital significance of checking immunization viability and proposing that sponsor dosages may ultimately be expected to reestablish the undeniable degrees of insurance noticed right off the bat in the inoculation program. These elements are essential to control the uplifted transmission of the delta variation as the population enters the impending pre-winter and winter viral respiratory season.

Notion and Attitude of People Towards the Vaccines and the Impact of That on Their Mental Health

Governments and health agencies have been at the forefront of fighting the pandemic, and one of the most incredible moves was manufacturing the vaccines. When people heard the messages of new vaccines available, more hopes were raised, as they were aware the pandemic would slow down if not end. However, the shock came when these vaccines were confirmed to be less than 100% efficient, making people more skeptical. With vaccines, the high burden of the effects of the virus globally is expected to be reduced. However, the public is quite hesitant about vaccination. The reasons behind this are still hard to understand, as many people are worried about the outcomes based on the effects of the vaccine that they hear and see on social media. This is a significant concern to the healthcare authorities, as they are worried about the progress of the pandemic if people are not taking the advisable precautions. Since the vaccines are available, there is little information regarding the acceptability and attitudes of the public toward vaccines for COVID-19.

The ground system sought after by most countries from one side of the world to the next was to lessen the compelling nature of the trouble, routinely by NonPharmaceuticals Interventions (NPIs). These include staying aware of the cover strategy, hand purging, social clearing out, travel obstructions, school terminations, and lacking or complete lockdowns (Manjula Bai, 2020). Presently, NPIs have the decision to control the improvement of the degradation. However, the most encouraging method to limit the pandemic and give yearning to lessen the mortality and awfulness rates stays inside the obstacle of clinical development. Such clinical improvement consolidates solid, safe, and reasonable antiviral organized trained professionals and immunizations. As of December 2020, no antiviral fixes were explicitly forefront against SARS-CoV-2 (Beigel et al., 2020). The US Food and Drug Administration (FDA) has allowed Remdesivir an Emergency Use Authorization for, indeed, not for the most part, talking around hospitalized patients with COVID-19 (U.S. Food and Drug Administration (FDA), 2020). Regardless, the WHO reprimanded its utilization in November 2020.

There is hope that the new vaccines will manage the pandemic, as people will have herd immunity (Omer et al., 2020). The vaccines are expected to raise herd immunity by 60-67% (Omer et al., 2020). However, this plan cannot be achieved because people are skeptical and hesitant to accept the vaccines. MacDonald and the World Health Organization Strategic Advisory Group of Experts (SAGE) (2015) defined vaccine hesitancy as “delay in acceptance or refusal of vaccination despite the availability of vaccination services” (p. 4161). The ability to accept vaccines is influenced by confidence, convenience, and complacency (Al-Mohaithef & Padhi, 2020). Conviction entails trust in the processes involved in the vaccines’ manufacture, safety, and effectiveness. Also, it involves trusting the authorities responsible for policymaking. As seen in most cases, many people do not trust the vaccination process, partly explaining why they are hesitant to be vaccinated. The safety of these vaccines is still questionable, making many people feel as if there are some ill intentions. Therefore, the government and concerned agencies must ensure that they make people feel safe. It will be a big challenge to vaccinate the whole population without this.

Moreover, the convenience of vaccination and relative ease of access to the vaccine, including physical availability, affordability, and accessibility, should be considered (MacDonald & SAGE, 2015, p. 4162). Vaccine complacency is associated with a low realized risk of vaccine-preventable disease and hence more negative attitudes toward vaccines (MacDonald & SAGE, 2015, p. 4162). Such distrust was shown in a survey directed in the US, where half of the Americans said they would take the immunization, 30% were uncertain, while 20% rejected the antibody (Neergaard and Fingerhut, 2020). In one more overview of grown-up Americans, 58% planned to be inoculated, 32% did not know, and 11% did not mean to be immunized (Fisher et al., 2020). Be that as it may, another review detailed that 67% of Americans would acknowledge a COVID-19 antibody, assuming it is prescribed to them, even though there was colossal segment contrasts in immunization acknowledgment (Malik et al., 2020).

Recognizing those mental characteristics might assist well-being authorities in persuading the sizable minority of Americans who do not need a COVID antibody. Sharing data has not worked. They were neglecting to convince an adequate number of individuals to get immunized, possibly missing the mark concerning the country’s objective of crowd insusceptibility when the whole populace would be safeguarded. Utilizing an idea from social brain research—the possibility that a bit of arrangement of moral instincts frames the establishments after which complex ethical perspectives are built—and applying it to an investigation of antibody wariness can assist with understanding this further. An unmistakable arrangement of mental characteristics offers another focal point through which to get wariness—and possibly new devices for general well-being authorities scrambling to convince individuals to get inoculated is the resultant reason. Cynics were significantly more likely than nonskeptics to have an exceptionally evolved awareness of freedom—the privileges of people—and to have the minor concession to those in important, influential places. Doubters were likewise two times as prone to care very much about their bodies and psyches’ virtue. They dislike things they consider disgusting. The outlook opposes slick order: It could be strict—halal or legitimate—or mainstream, like individuals who care profoundly about poisons in food varieties or the climate. Researchers have tracked comparative examples among cynics in Australia and Israel and an exhaustive sample of antibody-reluctant individuals in 24 nations in 2018 (Rossen et al., 2019; Kugler and Thaler, 2021). From the above argument, it can be argued that skepticism toward the

COVID-19 vaccine is highly viewed as a knowledge problem and is generally about gut beliefs. Therefore, developing ways to help overcome the negativity in their minds is essential.

Relative to mental wellness, a study by GarciaLlana et al. (2022) found that COVID-19 vaccination improved psychological distress (anxiety and depression scores) in chronic kidney disease patients. The three primary feelings of trepidation revealed by the members at standard were unfriendly impacts, not getting the immunization, and absence of data. This follows the theory that vaccination has a passionate defensive effect against pressure and the genuine dangers of the worldwide circumstance of isolation and the pandemic, as archived in different investigations (Strand & Parker, 2012).

Moreover, renal patients report that customized, well-disposed, and close correspondence of refreshed data conveyed by proficient medical care groups assisted them with adapting. The tracked-down information in the review, notwithstanding that from past writing, supports the significance of a multidisciplinary renal group, restorative correspondence, and consistent reassurance procedures as ways of assisting patients with managing the pressure gotten from having a constant ailment by further developing patient well-being results and advancing prosperity (Strand & Parker, 2012). It is essential to note that well-being goals in complex ongoing patients are not exclusively founded on estimating natural boundaries, such as creatinine, glomerular filtration, and safeguarding body capacities overall. They are likewise worried about advancing the patient's prosperity, which envelopes both enthusiastic and natural elements of a patient's well-being.

Low immunization acknowledgment among everyone is a not-kidding danger to fruitful COVID-19 inoculation, as a past examination has shown (Arshad et al., 2021; El-Elimat et al., 2021). Luckily, Spain has a high acknowledgment level of immunization among European nations. Accordingly, this variable may have affected kidney patients' view of immunization as a passionate asset rather than a danger. The mental effect of the circumstance made by the presence of new SARSCov-2 changes, the requirement for a good portion of immunization, and the impact of this large number of elements on the visualization of patients with chronic kidney disease still need to be characterized. New examinations will be expected to evaluate the effect of extensive antibody inclusion and other mental products. These outcomes are a boost for additional thought on the differentiated multidimensionality of the individual and the worth of collaboration. They likewise underscore the tremendous abundance of joint effort and participation during pandemic seasons. So, they mirror the advantages proposed to renal patients of such a methodology and the almost certain more noteworthy fulfillment it prizes to medical services experts.

Vaccination worked on mental pain in renal patients with clinical scores. No critical contrasts were found in discouragement, nervousness, and HealthRelated Quality of Life (HRLQOL) scores among the various gatherings of patients when immunization was completed. The discoveries of the current review could furnish well-being specialists with an exact base from which to plan a methodology that better advances mental prosperity for future inoculation crusades.

Methodologies of Self-Care to Manage Mental Health Issues

These methodologies are critical to one's psychological and actual well-being and can assist one with assuming responsibility for their life. People must deal with their bodies and psyche and associate with others to help their emotional wellness.

Deal With the Body

Mayo Clinic (2020) opines that one needs to be careful concerning their actual well-being. Get sufficient rest. Head to sleep and get up at similar times every day. Stick near their regular rest wake plan, regardless of whether they remain at home. Take part in actual or dinary work as it can decrease tension and further develop temperament. Observe an action that incorporates development—for example, dance or exercise applications. Get outside, for instance, on a nature trail or your patio. Notably, it is crucial to practice good eating habits. Pick a consistent eating routine. Try not to stack up on low-quality nourishment and refined sugar. Limit caffeine, as it can bother pressure, nervousness, and rest issues. Arguably, keep away from tobacco, liquor, and medications. Assuming one smokes tobacco, or then again, if he or she vapes, they are at a higher gamble of lung infection. Since COVID-19 influences the lungs, the risk increments considerably more. Utilizing liquor to adapt can exacerbate the situation and diminish one's adapting abilities. Try not to consume medications to adapt, except if your PCP recommended prescriptions for you.

Moreover, Mayo Clinic (2020) suggests that people must avoid more time on the screens. Switch off electronic gadgets for quite a while every day, including 30 minutes to an hour before sleep time. Put forth a mindful attempt to invest less energy before a screen— T.V., tablet, P.C., and telephone. Unwinding and reenergizing are essential. Put away opportunities for yourself. Indeed, even a couple of moments of quiet time can be refreshing and help settle one brain and diminish tension. Many individuals benefit from profound breathing, jujitsu, yoga, care, or contemplation. Absorb an air pocket shower, pay attention to music, or peruse or pay attention to a book—anything that assists one with unwinding. Select a method that works for you, and practice it consistently.

Deal With the Brain

First, decrease pressure triggers by keeping sta ndard daily schedule. Keeping a standard everyday plan is vital to psychological well-being—as well as adhering to a typical sleep schedule, saving reliable times for suppers, washing and getting dressed, developing work or study timetables, and exercising. Likewise, put away opportunities for activities one appreciates. This consistency can cause him or her to feel more in charge (Mayo

Clinic, 2020).

Mayo Clinic, (2020) advices that people need to limit openness to news media. Steady news about COVID-19 from a wide range of media can elevate fears about the illness. Limit online media that might open you to bits of gossip and bogus data. Additionally, limit perusing, hearing, or watching other news. However, stay up with the latest on public and neighborhood suggestions. Search for dependable sources, like the U.S. Communities for Disease Control and Prevention (CDC) and the World Health Organization (WHO).

Remain occupied since sound interruptions can move one away from the pattern of negative considerations that feed uneasiness and gloom. Appreciate leisure activities that one can do at home, like perusing a book, writing in a diary, making a specialty, messing around, or preparing another supper (Mayo Clinic, 2020). Alternatively, recognize another task, or clear out that wardrobe. Doing something positive to oversee tension is a solid survival technique. One is discursive, implying that zeroing in on sure contemplations and deciding to zero in on the good things in ones' day-to-day existence, rather than harping on how awful they feel, may be helpful. Consider beginning every day by posting something they are appreciative of. Keep a feeling of trust, acknowledge changes as they happen, and attempt to keep issues in context.

Utilize your ethical compass or profound life for help. Assuming you draw strength from a conviction framework can bring you solace during troublesome and dubious times. Put forth boundaries (Mayo Clinic, 2020). Try not to become overpowered by making a groundbreaking rundown of things to accomplish at home. Put forth sensible objectives every day, and blueprint steps you can take to arrive at those objectives. Give yourself credit for each positive development, regardless of how little. Also, perceive that a few days will be preferable over others.

Interface With Others

Construct support and reinforce connections by making associations. Avoid social seclusion when working from home or disconnecting with others because of COVID-19. Furthermore, one should figure out opportunities every day to make virtual associations by email, messages, telephone, or video talk. Assuming that they are working from home, they should ask their associates how they are doing, and share adapting tips. Appreciate virtual mingling and conversing with those in your home. On the off chance that they are not wholly inoculated, one need to be inventive and safe. In contrast, interfacing with others' faces, taking strolls, visiting the carport, and other outside exercises necessitate wearing a veil for indoor activities.

Assuming that they are entirely vaccinated, they can even securely return to numerous indoor and open-air exercises they might not have had the option to do during the pandemic, like social affairs with loved ones (Mayo Clinic, 2020). For unvaccinated individuals, open-air exercises that permit much room among you and others represent a lower hazard of the spread of the COVID-19 infection than indoor exercises do. Notwithstanding, assuming one is in a space with countless new COVID-19 cases somewhat recently, the CDC suggests wearing a cover inside in broad daylight, outside in jam-packed regions, or in close contact with unvaccinated individuals.

It is noteworthy to work on something for other people. Thus, one should track down the reason for aiding individuals around them. Helping other people is a magnificent method for assisting ourselves. For instance, email, text, or call to watch their companions, relatives, and neighbors—particularly the more seasoned individuals. Suppose you know somebody who cannot get out: inquire whether something is required, such as food or a solution. It is highly advisable to support a relative or companion (Mayo Clinic, 2020). Think of ways to keep in touch on the off chance a relative or companion should isolate at home or in the medical clinic because of COVID-19. For instance, it could be through electronic gadgets, on the phone, or by sending a note to light up the day.

Keep Away from Shame and Separation

Mayo Clinic (2020) warns that shame can cause individuals to feel disengaged and surprisingly deserted. They might feel discouraged, hurt, and furious when companions and others locally stay away from them, inspired by a paranoid fear of getting COVID-19.

Disgrace hurts individuals' well-being and prosperity in numerous ways. Criticized gatherings may frequently be denied the assets they need to focus on themselves and their families during a pandemic. Moreover, individuals stressed over being trashed might be more averse to getting clinical consideration. Individuals who are derided might be prohibited or disregarded; treated unexpectedly; denied work and instructive open doors; and focused on verbal, passionate, and actual maltreatment. Those who have encountered disgrace connected with COVID-19 include individuals of Asian descent, medical services laborers, individuals with COVID-19, and those set free from isolation.

One can decrease disgrace by taking action, such as:

- Getting current realities about COVID-19 from trustworthy sources like the CDC and WHO
- Making some noise on the off chance that they hear or see wrong explanations about COVID-19 and specific individuals or gatherings
- Contacting individuals who feel criticized
- Showing support for medical services to laborers

Perceive What Run of The Mill and So Forth Is

Stress is a normal mental and actual response to the requests of life. Everybody responds contrastingly to tough spots, and feeling pressure and stress during an emergency is typical. In any case, various difficulties—for example, the impacts of the COVID-19 pandemic— can push one past his/her capacity to adapt. Many individuals might have emotional well-being concerns, like indications of uneasiness and sadness during this time. Moreover, sentiments might change after some time (Mayo Clinic, 2020). Despite one's earnest attempts, one might wind up feeling powerless, miserable, angry, bad-tempered, sad, restless, or apprehensive. One might have trouble focusing on commonplace undertakings, changes in hunger, body throbbing painfulness or trouble dozing, or might battle to confront routine tasks. At the point when these signs and manifestations keep going for a long time, make one hopeless, and bring on some issues in their day-to-day existence so they find it difficult to complete ordinary obligations, it is an ideal opportunity to request help.

Get Assistance When One Requires It

Assuming one has concerns, or again, on the off chance that they experience worsening emotional wellness manifestations, they should request assistance when required and be forthright about how they are doing. Trusting emotional wellness issues, such as tension or discouragement, will disappear alone and prompt deterioration. To get the help, one might need to:

1. Call or utilize web-based media to contact a dear companion or adored one—even though it could be challenging to discuss one's sentiments.
2. Contact a clergyperson, otherworldly pioneer, or somebody in their confident local area.
3. Contact one's worker help program, assuming their manager has one, and request guidance or a reference to emotional wellness proficiency.
4. Call an essential consideration supplier or emotional wellness expert to get information about arrangement choices, discuss tension or sorrow, and get counsel and direction. Some might give the option of telephone, video, or online arrangements.
5. Contact associations like the National Alliance on Mental Illness (NAMI), the Substance Abuse and Mental Health Services Administration (SAMHSA), or the Anxiety and Depression Association of America for help and direction on data and treatment choices.

If they feel self-destructive or consider harming themselves, they should seek help. They can contact their essential consideration supplier or psychological wellness proficient. Alternatively, call a self-destruction hotline or call the National Suicide Prevention Lifeline at 1-800-273-TALK (1-800-2738255).

Finally, proceed with taking care of oneself procedures. People might anticipate that their overwhelming inclinations should blur when the pandemic ends. However, stress will not vanish from their life when the well-being emergency of COVID-19 closes. Proceed with these taking care of oneself practices to deal with one's emotional well-being and increment their capacity to adapt to life's ongoing difficulties (Mayo Clinic, 2020).

Applying Cognitive Dissonance to Increase Compliance with Preventive Measures and Reducing the Psychological Effects

People may believe in reducing or preventing viral acquisition, but they will never take the needed action. Failure to take the necessary steps exposes them to the infection and views the whole process as a scam. Therefore, one way to reduce the pandemic's psychological effect is by increasing compliance with the set safety measures (Cancino-Montecinos et al., 2020). When people comply with the standards, they will be safe, and that safety will, in turn, increase their mental health stability. Cognitive dissonance is an effective tool that can be applied to this.

Simultaneously, long coronavirus is another illness characterized as it is dealt with and made. The individual, populace, and financial weight have been underplayed because of an emphasis on intense hospitalization and mortality (Cancino-Montecinos et al., 2020). Conventional proof exists in some opportunity to collect data, featuring the significance of patients' insight and the ability of forefront clinicians in arrangement and framework reactions. In any case, this is not reflected in the enrollment of warning councils, dynamic associations, or examination subsidizing calls that have focused on instrument and pathophysiology above the plan and assessment of clinical administrations. This makes possible snags in acknowledgment and attention to long coronavirus and deferred determination and evaluation of treatment techniques (Cancino-Montecinos et al., 2020). Adjusting how patient and public inclusion is seen and sanctioned is also very much past due. The way forward should include acknowledging impediments of ebb and flow survey boards and administration constructions to manage the scale of these new sicknesses and empower more considerable cross-fertilization between research, clinical, general well-being, and strategy circles.

As of December 18, 2020, COVID sickness has 72 million affirmed cases and has prompted 1.6 million deaths worldwide. Strategies and rules have been developed to safeguard general well-being (Mallapaty, 2021). However, such approaches are not restricted to mask-wearing, closing enormous social affairs, and limiting or dealing with the progression of distinct regions. Individuals have been urged to embrace COVIDpreventive practices (CPP) by rehearsing social separation, wearing a veil out in the open spots, taking part in self-disconnection if necessary, and expanding the recurrent term of hand washing. The viability of the previously mentioned COVID control strategies has been inspected in late examinations.

To assess the effect of obligatory cover-wearing in broad daylight spaces, Mitze et al. (2020) set the city of Jena, Germany, as a kind of perspective and contrasted the COVID measurements there and those of different urban areas like Jena; however, they studied measures without the compulsory veil wearing arrangement. The result showed that obligatory cover-wearing diminished the day-by-day development pace of COVID-19 in Jena by 40%, in this manner uncovering the viability of covers in forestalling COVID-contaminations. In expansion, Eikenberry et al. (2020) surveyed the impact of veil-wearing on a local area by fostering a transmission model utilizing the information connected with COVID-19 elements in New York and Washington. The outcomes demonstrated that mask-wearing safeguards sound people and forestall asymptomatic transmission. To assess the connection between non-drug medications (for example, shutting borders, having quarantine and detachment, and rehearsing CPP) and the COVID-19 contagiousness rate, Cowling et al. (2020) measurably investigated the contamination rate. Hong Kong also noticed a negative relationship between non-drug mediation and the COVID-19 contagiousness rate.

Tang et al. (2020) developed a period subordinate unique model in Wuhan, China. The creators proposed that diligent and severe self-isolation plays a significant part in lessening the plague transmission speed. Additionally, Chu et al. (2020) performed a systematic audit and meta-examination, which inspected the impact of social removal, veil-wearing, and eye security on the forestalling individual-to-individual transmission of the severe, intense respiratory disorder COVID-19. The outcomes showed that such practices could successfully diminish the disease pace of COVID-19.

In the period of online media, open science, and consistent patterns of media reporting, citizens generally view different nations and settings as much as their own. Given news and examination from China and Italy in mid-2020, one could contend that the primary UK lockdown was postponed (Tang et al., 2020). However, the subsequent lockdown was most certainly late decided to the extended effect of Coronavirus on high-hazard populaces, resulting in an abundance of mortality and indirect consequences for non-Coronavirus care. Ways to deal with covers, the opening of schools and colleges, and line controls are different occurrences. There has been a distinction between distributed proof or strategy in other nations and the UK circumstance. The inevitable choice to inoculate youngsters in the UK happened half a month after school began when cases had started to rise. A few different nations had effectively created immunization programs for those beyond 12- and 16 years old with distributed information. Deferrals contrasts, or errors in well-being strategy and the reasoning for independent directors should be conveyed to people.

The consistent theme is the hole in transparent, public correspondence of data and affirmation of the impediments and public association in an independent direction. This structure might assist the public, scientists, well-being experts, and policymakers navigate pressures in the quickly moving proof and strategy scene of crisis readiness. Mental disharmony might be logical or even inescapable in pandemics. However, decreased systems can incorporate procuring new data; evolving convictions, perspectives, or practices; or diminishing the significance of the noisy evolving beliefs, mentalities, or conventions.

A Call to Hold Cognitive Dissonance Based on Personal Experience- Elaborative Story

As the pandemic hit, a companion asked me, "What has your experience of this time been like?" I felt constrained to react with something about how awful it has been because I should feel. This has been a period of languishing over many, especially for individuals who currently live in emergency constantly. It has featured holes in medical care, with departing individuals at the bleeding edges unprotected from the infection. It has featured the gaps in race and destitution, leaving principally dark and Latinx populations in places with inadequate security. There is no choice to remove it ("Social Distancing is a Privilege"). Also, valuable individuals are dying, both from the infection and from not having the option to get the consideration they need since everybody is extended further. Fear is increasingly up, and individuals are skeptical. The rundown is unending. However, simultaneously, this is not my insight (yet). Truth be told, as I told my companion, I am holding a great deal of mental disharmony during this time. Mental disharmony is the point at which we have disconnected convictions, which is awkward for us, and we want to determine the cacophony through different means, including limiting one side and choosing without a doubt—for this situation, this means choosing regardless of whether these COVID-19 rules have been a great or a terrible experience.

Here is the dissonance: while there are times when I want to see more individuals in 3D, and there are loved ones I am stressed over, all in all, I am encountering the next splendid minutes. As somebody who routinely goes to work, I am more than happy to find myself dependably ready to make week-after-week responsibilities. My schedules are ordinary, and I contemplate more. I practice more. The multiplication of online classes takes care of my need to learn new things without adding to my drive time. I am resting for more hours. In any event, I am associating even more routinely with dear companions who do not live near me, finding new virtual exercises we most likely might have done pre-COVID-19; however, I did not. I am a social, contemplative person who routinely overschedules herself, and this new reality has constrained me into balance, considering the restoration I want. There are minutes when I fear a re-visitation of my pre-COVID-19 life. What is more, this is because I live with a great deal of honor, have been fortunate, and have a considerable rundown of things to be appreciative of; both items can be valid—that this is an awful time and a fantastic time.

We gab about the significance of seeing the whole picture at our workplace, which expects us to hold cognitive dissonance. Individuals are a blend of good and terrible. Conduct might have both great and awful impacts on us. To build admittance to prosperity for everybody, we need to battle against just searching out data that affirms what we need to see (the tendency to look for predictable feedback), and we need to search for data that conflicts with our impressions purposely. We need to battle against the inclination to disregard what may disagree with what we accept or experience to be valid.

When we shift our viewpoints to appreciate the situation entirely, we have a more profound comprehension of the tradeoffs we are, for the most part, gauging and what is driving conduct. It is discord that permits us to perceive the complicated and untidy ways we are mainly carrying on with our lives and the trouble that change brings—even change that is comprehensively considered great. It permits us to have essential multifaceted solutions to why that individual leaves that relationship. For what reason do they do that, regardless of whether they realize it harms them? For what reason could I stop (fill in the blank)? Also, along these lines, it permits us to have more successful complex strides forward.

Individuals' admittance to prosperity relies upon our capacity to hold various facts. It expects us to neutralize how our minds are wired, to incline toward distress. It is a troublesome and significant spot to be. Along these lines, I am working on doing precisely that. I am looking for data that does not squeeze into my experience, and I am giving my best to perceive, approve, and react to the two fundamental factors of this time—and more as I find them—while lessening neither of them (DeTienne et al., 2020).

The above experience is an excellent lesson on how cognitive dissonance can be applied in helping to curb the pandemic by allowing people to do what is right to keep them safe. The right thing may be against a person's wish, but the person benefits and becomes free from the infections at the end of it all. Therefore, through the application of cognitive dissonance, a person will develop negative emotions, which the individual will be motivated to reduce, and similar situations will probably be avoided in the future.

Chapter 3: History of Pandemics, Lessons, and there Psychological Impacts

---

Human history is not complete without the mention of pandemics. For an extended period, pandemics have robbed humans of great opportunities and have consistently demonstrated how poorly structured human systems are. Like the recent COVID-19 pandemic, past pandemics have greatly affected people psychologically, and until today, studying the effects still brings disturbing memories (Mallapaty, 2021). In the domain of irresistible sicknesses, a pandemic is an imaginable direct outcome. The point when a pestilence spreads past a nation's lines is the point at which the disease formally turns into a pandemic. Contagious illnesses existed during humanity's agrarian days; however, the shift to agricultural life 10,000 years prior made networks that made pestilences more conceivable. Jungle fever, tuberculosis, infection, flu, smallpox, and others previously showed up during this period (Littman, 2009). The more acculturated people became, building urban areas and fashioning shipping lanes to associate with different urban areas and taking up arms with them, the more potential pandemics became. Understanding the psychological impact of past pandemics and how they were managed contributes many lessons on how the current pandemic may be addressed. When keenly studied, there is a close relationship between the pandemics, which brings about a point of reference applicable in management.

The Pandemics

The Plague of Athens

Of the historical human pandemics, the Plague of Athens is the oldest, and it mainly affected the city of Athens in 460 B.C. during the Peloponnesian War. It claimed almost two-thirds of the city population. While the disease did not originate from the city, it was realized after affecting cities like Libya, Egypt, and Ethiopia (Littman, 2009). The manifestations included fever, thirst, ridiculous throat and tongue, red skin, and sores. The illness, suspected to have been typhoid fever, debilitated the Athenians essentially and was a huge component in their loss to the Spartans (Littman, 2009).

The Antonine Plague (165 AD)

The Antonine plague was perhaps an early appearance of smallpox that started with the Huns. The Huns then, at that point, tainted the Germans, who passed it to the Romans, and afterward, returning soldiers spread it all through the Roman domain (Scheidel, 2009). Side effects included fever, sore throat, runs, and discharge-filled bruises if the patient lived sufficiently long (Scheidel, 2009. This plague proceeded until around 180 A.D., asserting Emperor Marcus Aurelius as one of its casualties. Many historians believed this plague was the reason behind the fall of the Roman Empire, as it claimed almost 5 million lives.

250 A.D.: Cyprian Plague

Named after Carthage's main known casualty, the Christian diocesan, the Cyprian plague involved diarrhea, regurgitating, throat ulcers, fever, and gangrenous hands and feet. City occupants escaped to the country to avoid contamination and spread the infection further. Conceivably beginning in Ethiopia, it went through northern Africa, into Rome, then, at that point, onto Egypt and toward the north.

541 A.D.: Justinian Plague

It started in Egypt and killed almost 26% of the world's population.

11th Century Leprosy

Presently known as Hansen's sickness, it besets many individuals for a year and can be deadly if not treated with anti-microbial. A sluggish, creating bacterial sickness that causes bruises and deformities, the disease was accepted as a discipline from God that ran in families (Littman, 2009). This conviction prompted moral decisions and the exclusion of casualties.

Black Death

Like the Plague of Justinian, the bubonic plague brought the Black Death. The quick spread of the sickness keeps astounding students of history and disease transmission specialists (Lotz-Heumann, 2020). The focal clarification exists in trademark highlights of middle-aged society in a unique period of modernization, proclaiming the change from an archaic to early present-day European culture (Lotz-Heumann, 2020). Huge boats conveyed products across consistently broadening, delivering courses throughout Europe and then some. This framework for significant distance exchange was enhanced by a snare of animated short and medium distance conversation that bound together populaces around the Old World, naming this the brilliant time of microbes.

1492: The Columbian Exchange

Mercantilism, a financial hypothesis that dismissed streamlined commerce and advanced unofficial law of the economy to upgrade state power, characterized the monetary strategy of European colonizing nations (Khan Academy, 2021). Christopher Columbus presented ponies, sugar plants, and infections to the New World while presenting items like sugar, tobacco, chocolate, and potatoes to the Old World. The cycle by which items, individuals, and illnesses crossed the Atlantic is known as the Columbian Exchange (Khan Academy, 2021).

1665: The Great Plague of London

The Black Death's last significant episode in Great Britain caused a mass departure from London, driven by King Charles II. The plague began in April 1665 and spread quickly through the blistering midyear months. Bugs from plague-contaminated rodents were one of the primary transmission drivers (Jarus, 2020). When the plague finished, around 100,000 individuals, including

15% of the inhabitants in London had died. Nonetheless, this was not the finish of that city's misery (Jarus, 2020). On September 2, 1666, the Great Fire of London began for four days and torched an enormous part of the city.

1817-1923: The Cholera Pandemics

The primary cholera pandemic arose out of the Ganges Delta with an episode in Jessore, India, in 1817, coming from tainted rice. The illness immediately spread throughout India, Myanmar, and cutting-edge Sri Lanka by going along shipping lanes laid out by Europeans. By 1820, cholera had spread to Thailand, Indonesia (killing 100,000 individuals on Java alone), and the Philippines. The sickness advanced toward China in 1820 and Japan in 1822 via tainted individuals on ships from Thailand and Indonesia. It additionally spread past Asia. 1821 British soldiers from India to Oman carried cholera to the Persian Gulf. The infection, at last, advanced toward the European domain, arriving at current Turkey, Syria, and southern Russia. The pandemic vanished six years after it started because of an extreme winter in 18231824, which might have killed the microorganisms living in water supplies.

The Late 1800s: Yellow Fever

Yellow fever pestilence struck the United States over and over in the 18th and 19th centuries. The sickness was not native; pandemics were imported by transport from the Caribbean. Before 1822, yellow fever assaulted urban communities as far north as Boston; however, after 1822, it was limited to the south (Jarus, 2020).

The 1918 Flu

The coronavirus pandemic has enlivened bunches of correlations with the 1918 influenza, now and again called Spanish flu, which got its name not because it began in Spain but because of World War I. Spain was the central nation speaking the truth about the pandemic's cost. Influenza came in two waves, starting in 1918 and finishing in 1920. The quantity of contaminated is faltering—upwards of 500 million, with assessments of 50 million deaths around the world. Detachment and strategic isolations were utilized to dial back transmission. Indeed, even President Woodrow Wilson was blasted, almost wrecking talks at the Paris Peace Conference, where influenza left him laid up for quite a long time.

1957-1958 Asian Flu

The 1957-1958 Asian influenza pandemic was a worldwide pandemic of flu. It was an infection subtype H2N2 that started in Guizhou in southern China. The number of overabundant passings brought about by the pandemic is assessed to be 1-4 million all over the planet (1957-1958 and likely past), making it probably the deadliest pandemic ever (Pineo, 2021). After 10 years, a reassorted viral strain, H3N2, caused the Hong Kong influenza pandemic (1968-1969). The economic effects of this pandemic were significant. The Dow Jones Industrial Average lost 15% of its worth in the final part of 1957, and the U.S. encountered a recession. In the U.K., the public authority paid £10,000,000 in affliction advantage, and a few production lines and mines needed to close. Many schools were required to shut in Ireland, remembering 17 for Dublin. The reason for the significant pandemic effects and potential lessons from this pandemic is that the preparedness rate was low. Nobody saw the results surmounting what it came to at the end.

1981 HIV/AIDS Pandemic

Those contaminated by the HIV infection experience fever, cerebral pain, and broadened lymph hubs upon disease. Transporters become exceptionally irresistible through blood and genital liquid whenever indications die, and the illness annihilates lymphocytes. First distinguished after quite a while, it destroys an individual's invulnerable framework, bringing about inevitable passing by sicknesses that the body would generally ward off (Beyrer, 2021). HIV was first seen in American gay networks, yet it is accepted to have been created from a chimpanzee infection from West Africa during the 1920s (Beyrer, 2021). The condition, which spreads through specific body liquids, moved to Haiti during the 1960s and New York and San Francisco during the 1970s. Medicines have been created to slow the advancement of sickness. However, 35 million individuals worldwide have passed from AIDS since its disclosure, and a fix is yet to be found (Beyrer, 2021).

2009 Swine Flu

Before the coronavirus, this was the world's latest pandemic, contaminating as much as 21% of the total populace. The influenza was a mixed bag of several strains that had never been aggregately seen together (Carvalho et al., 2020). Most of those contaminated by the influenza were kids and youthful grown-ups, with more established individuals— most likely to die from this season's virus— invulnerable to it now (Carvalho et al., 2020).

Lessons From the Past Pandemics

The past pandemic is an excellent reference source for managing the current pandemic, especially mental health. The current COVID-19 pandemic is even more easily manageable since the advancement in healthcare has provided great inspiration and opportunities to help manage it (Saladino et al., 2020). From past pandemic examinations, mental perseverance is a test many individuals face. For psychological well-being, elevated pressure reactions to the pandemic appear in three ways: a) advancement of another episode of an issue in those with an inclination to a significant mental issue or an intense intensification in the individuals who, as of now, have such a turmoil, b) improvement of an injury or stressor-related confusion—for example, intense pressure issues, Post Traumatic Stress Disorder (PTSD), or change problems, and c) improvement of an indicative pressure reaction that does not meet the symptomatic measures of a mental issue. Excessive use of alcohol, PTSD, nervousness, outrage, anxiety toward disease, vulnerability, and doubt are some of the pandemics' prompt and long-haul impacts. Distinguishing individuals needing emotional well-being care and deciding the proper mental administration and treatment required will be significant. Expanding the utilization and accessibility of telehealth, bunch gatherings, and online assets are a few different ways medical care laborers can plan for the expanding request of mental administrations during and following the pandemic.

What Do the Lessons Imply for The Future Management of Mental Health Issues?

Many lessons can be learned from past pandemics. These pandemics—including the flare-up of SARS CoV-1 out in 2003, 2014 Ebola, 2015 MERS, 2001 Anthrax, reactions to past cataclysmic events, and previously distributed articles on COVID-19—propose the potential psychological impact during and following the COVID-19 pandemic. Liquor use issues, PTSD, tension, outrage, anxiety toward infection, saw hazard, vulnerability, and doubt are only some of the quick and long-haul impacts that will probably result from the COVID-19 pandemic (Saladino et al., 2020). Moreover, long-haul well-being hazard practices, trouble responses, and improper survival techniques represent a danger of intensifying mental issues and compromising mental strength.

Patients with existing psychological sicknesses, like schizophrenia, sadness, uneasiness, and bipolar problems, will probably struggle to adapt to the impacts of a pandemic (Esterwood & Saeed, 2020). Likewise, there is an expanded danger of intensification in patients who might be steady on support therapies. The stressors that have been seen after previous seasons of emergency represent a troublesome obstruction for patients to survive. People with a characteristic weakness to foster a psychological problem address another weak populace. The stressors presented by the COVID-19 pandemic might undermine their psychological security, indicating illnesses like PTSD, melancholy, new-beginning psychosis, tension, or substance use issues. In conclusion, people like medical care laborers, relatives of contaminated people, and those isolated presented a strange measure of pressure to address a weak populace (Galea et al., 2020). While this last gathering may not foster actions for a psychological problem, they are conceivably in danger of requiring mental administrations to conquer the over-the-top measurement of pressure they have encountered.

Change in Approach

The drawn-out effect of COVID-19 and other pandemics on emotional wellness and prosperity will probably require a long time before it turns out to be completely obvious (Galea et al., 2020). Meanwhile, dealing with this effect will require a coordinated exertion from the medical care framework overall, not simply from psychological well-being care suppliers. It will be vital to recognize patients with existing diseases in extreme emergencies and new instances of psychological sickness. For people not recently analyzed, offer help for the individuals who do not meet rules for a psychological issue but will require treatment. An expanded screening measure should be performed for these three gatherings to be distinguished and for benefits to be accessible.

The proper mental administration and treatment should be custom fitted to introduce issues (Galea et al., 2020). To give some examples, this necessitates instruction for strategies for dealing with challenging times, stress transformation, social and mental treatment, and pharmacotherapy. Pharmacotherapy should be observed for antagonistic impacts and medication cooperation with this flood in cognitive issues. Patients should be surveyed satisfactorily for treatment-based administrations to recognize which treatments are shown and accessible (Galea et al., 2020).

There is a requirement for additional exploration to survey the effect of the pandemic. The plan or distinguishing instruments can evaluate for shifted introductions, as discussed above, to decide on successful and proficient consideration conveyance models and assess the effect of COVID-19 on vulnerable populations. For example, experts must care for youngsters and teenagers, individuals who face obstructions in getting to medical services, and those having a place with lower financial layers. For people who do not meet models for a clinical finding, methods for dealing with hardship or stress, backing, and assets ought to be given. A different line of exploration will be expected to decide the psychological well-being ramifications of social disengagement for weak gatherings and how these could be relieved under pandemic conditions. Moreover, the neuropsychiatric effect of COVID-19 is neglected to a great extent. Research that assesses the direct neuropsychiatric outcomes will likewise be expected to further develop the treatment, psychological wellness care, and preventive measures during possible ensuing pandemics (Troyer et al., 2020).

Need to Safeguarding Mental Health and well-being in an Already Saturated Area of Health Care

All pandemics significantly affect people's psychological well-being (Galea et al., 2020). As opposed to past pandemics, the COVID-19 pandemic has indicated the most extraordinary lapse and hole in the healthcare system, especially when it comes to the management of mental health. Most healthcare systems are already complete or overwhelmed, and there is a need to safeguard the mental health departments lest the population lose it.

With psychiatrists' public labor force deficiency— an essential consideration for doctors and emotional wellness suppliers before COVID-19—it is necessary to discuss how the medical care field will address the expanding number of patients needing mental administration (Esterwood & Saeed, 2020). Through telehealth, emotional well-being care suppliers can oblige an expanding interest in psychological well-being administrations. With social separating rules, telehealth has become a considerably more utilized asset, acquiring commonality among suppliers, doctors, and patients (Esterwood & Saeed, 2020). With a set number of emotional well-being care experts and a developing number of individuals needing administration, bunch treatment may likewise end up being helpful.

Notwithstanding telehealth and gathering treatment, online assets will be significant for patients and people needing emotional well-being assets (Esterwood & Saeed, 2020). Mental, social therapy, reflection, and other web-based modules that individuals can do at home individually could be valuable for those who cannot plan an eye-to-eye meeting with a specialist, therapist, or other psychological wellness suppliers. During the pandemic, people will likewise have to track down ways of making physical and social backings to guarantee psychological wellness during these seasons of social removal (Esterwood & Saeed, 2020). Likewise, The pandemic gives people a valuable chance to foster populace-level ways to give avoidance-centered and proof-based emotional well-being care that is proficient and powerful and rise out of this pandemic with better approaches. People should perceive that the worldwide COVID-19 pandemic will probably be trailed by one more pandemic of mental and conduct ailments, and everyone wants to carry out the means expected to relieve it. Clean correspondence and secure administrations from a multidisciplinary psychological wellness care group should be laid out to oversee rising mental requests satisfactorily. The field of medication is being tested in uncommon ways, requiring change. The encounters of COVID-19 pandemic will influence every person indefinitely in various ways and will have enduring impacts. It is essential to resolve the issue of emotional wellness during and after this pandemic.

Chapter 4: Impact of Misinformation Epidemic During COVID-19 Pandemic on Mental Wellness

Apart from the challenges caused by the COVID-19 pandemic on healthcare all over, it has been a good platform for spreading rumors, misinformation, hoaxes, and unfounded data on the etiology, outcomes, prevention, and cure of the disease. People always seek data or information from all points whenever the pandemic strikes since relying on a single media may be risky. Since people are all over the internet seeking solutions to the conditions, the misinformation tends to cloud the healthy behaviors that ultimately contain the viral spread and ultimately result in poor physical and mental health outcomes among individuals. A single mishap of the information to the public has detrimental consequences. Several incidences have been reported across the globe that have resulted from such misinformation. Addressing this issue requires the concerned agencies and frontline healthcare providers to have valid and most recent scientific data to help curb the underlying effect of being misinformed. Mass media houses, healthcare stakeholders, and community-based organizations are challenged to have strategic and dependable partnerships to disseminate valid and accurate public health messages during these challenging times. Notably, advanced data processing technologies should be used to ascertain the validity of information reaching the public. Furthermore, these practices should be controlled with administrative and regulation implementation measures close by, guaranteeing telemedicine-put-together administrations giving exact data concerning COVID-19. Misinformation about healthcare data can significantly affect people's mental health, as indicated in the following sections.

Misinformation, Rumors, and Hoaxes and Their Impact

Since COVID-19 is a global pandemic, there are high chances of several hoaxes, pieces of misinformation, and rumors about it (Raman, 2020). People will give different versions of the story on the disease's spread, etiology, and prevention. The spread of fake news is high, and whenever it happens, the ecosystem of the news conveyance is broken, and people tend not to know what to believe and what not to believe. For instance, the 2016 USA elections culminated in fake news that spread faster than the news from 19 reliable sources. The spread of misleading information is of significant health concern since exposing an individual to a large volume of contradicting data will result in media fatigue, making the person discontinue the prescribed healthy habits. Moreover, they avoid healthy behaviors such as handwashing, wearing masks, and even social distancing when exposed to such erroneous and unfounded data. Instead, start practicing incorrect practices with the potential of increasing the viral spread and making them have poor physical and mental health outcomes.

Some examples of the fake news's impact on the general population include a father committing suicide in India after learning that he has COVID-19. The man left three kids fatherless and distressed about what life offers without a father. He watched the uploaded online videos showing how people were forced to quarantine in China and how people died from the COVID infection. Mr. Bala Krishna's family attempted to dissuade him, contending that he had not encountered any other individual experiencing the disease. Nonetheless, he self-isolated himself and pelted loved ones with stones when they drew closer. As reports in the Times of India indicate, he escaped the family home during the long early stretches of Monday and ended his own life. It is believed to be the principal passing by self-destruction anyplace on the planet connected to the COVID flare-up. Consequently, there was a reported overdose of chloroquine in Nigeria after Trump endorsed it as an alternative COVID-19 treatment (Busari & Adebayo, 2020). Apart from the surge in drug price by almost 400%, many people have abused this drug and overdosed themselves, thinking they would get healed or be immune from the disease.

Tales about a public lockdown in the U.S. enabled alarm purchasing of food and paper things, upset the stock association, and exacerbated request supply openings and food weakness among people with low financial status and other weak people groups (Tasnim et al., 2020). Comparative mannerisms have been seen in different economies, from one side of the world to another, which could prolongedly impact thriving general food and psychosocial success among the affected groups. The certain duplicity related to the finding and treatment of COVID-19 has baffled both everyone and clinical advantages suppliers, who are fighting all things considering under-investigated pollution (Tasnim et al., 2020). This peculiarity is besides diminishing the authenticity of new actual exposures regarding expected fixes or balancing specialist contenders for this issue (Tasnim et al.,

2020). These stunts and bits of clamor are comparatively causing social shame around COVID-19, which has accomplished reduced consistency with home quarantine and social division. Several nations have detailed episodes where numerous people were dirtied by a solitary individual who visited their mosque or church, no matter their fundamental thought specialist's proposition to stay home-bound (Tasnim et al., 2020).

Unvetted information is straightforwardly open through online media (Ahmed Siddiqui et al., 2020). Evaluation pieces are viewed as fundamental factors, even without any scientific foundations. A ceaseless stream of data on the pandemic has created a need to continue to move and strain. Repeated receptiveness to this flood of lies could impact outside this present reality (Ahmed Siddiqui et al., 2020). This could incite a fantasylike experience, which has been associated with anxiety and online media misuse. Social separation has affected the balance between associations and enthusiastic relationships between veritable and virtual (Ahmed Siddiqui et al., 2020). Without a doubt, the world is in a virtual, long stretch, genuinely blamed relationship of sorts with online media. This relationship has provoked a fantasy-like experience, impacting various people segregated from the real world, with online media as the common component.

It is legitimate that people living in segregation, withdrawal, or at risk for an overwhelming disease discharge are most likely to experience psychosocial stress and negative prosperity results, which could instigate their benefit in getting to know the affliction more (Tasnim et al., 2020). Regardless, such conditions require confirmation enhanced by a movement of the correct information. The WHO has perceived this need, which has helped a couple of online media stages and seven huge tech associations. Facebook, Google, LinkedIn, Microsoft, Reddit, Twitter, and YouTube agreed to dispose of deception and misrepresentation and propel essential updates from the clinical benefits workplace (Tasnim et al., 2020). Regardless of these plans, pieces of prattle are uncontrolled on a considerable number of stages, an issue that requires a couple of multipronged interventions to address.

Solving Misinformation and Its Effects

Public and Frontline Healthcare Providers' Education

The first victims of misinformation are the public, who directly receives the source's data. Therefore, the first approach is educating the people on credible data sources (Alonso-Galbán & Alemañy-Castilla, 2020).

These sources include the WHO, UNICEF, the Department of Health, government agencies, and the CDC. These organizations will not release any information unless verified and ascertained to be true. Public education and sensitization on credible sources of information are essential since they will have access to information relevant to their health. Secondly, the duty lies with the frontline healthcare providers (Alonso-Galbán & Alemañy-Castilla, 2020). They should ensure they rely on credible sources with scientific backgrounds. They should rely upon the most recent publications and sources of information to guide their care process for the infected or at-risk population. Doing this will help reduce the usage of wrong procedures that would affect the patient's health. Doing so may work with more overwhelming patient-provider correspondence and result in better prosperity results. Ahead of time, the coordination of information advancement successfully chipped away at lenient thought in managing mass mishap events, yielding a couple of positive outcomes, such as smoothing out of resources, an extension of care, and further creating prosperity consequences for patients (Zhao et al., 2006).

Using Mass Media, Community Organizations, Support Groups, and Civil Society to Disseminate Credible Information

Comprehensive correspondences, neighborhoods, support parties, and ordinary society could arguably acknowledge fundamental parts in dispersing cautious data. To accomplish this, it is essential to make the entire relationship at district and general levels, associating disengaged and online assets in a planned way so that upheld data is passed on across stages. Also, all stunts and reports should be shed from all online locations, and instrumental data should be permitted to further the best childcare for everybody. Best-in-class headways, like conventional language or text-mining approaches, may help perceive web content without a great explanation. The most recent information mining assessments effectively perceive the fantastic attributes of a vast extent of fake news and kill them from various stages. Twitter, Reddit, and Amazon have effectively finished these procedures to forgo counterfeit records or surveys. The use of practically identical rules to kill all bits of clamor and trickery concerning this pandemic is subsequently obviously reachable, regardless of its incredible hardships. This way, online media and other electronic suppliers should embrace such measures to perceive and dispose of possibly horrendous trickiness and stories.

Arresting Disseminators of False Data and Closing the Accounts

Furthermore, online entryways and faculty attracted to the creation and extension of such falsehood ought to be stopped (Ministry of Electronics & Information Technology, Government of India, 2021). Nearby and public administrative well-informed authorities and rule need affiliations should be mindful of these difficulties to decide these issues in their various settings. In a manner, the states of a colossal number of nations have gotten people related with the spread bits of information. In any case, to keep up with rules globally, specialists should lead a reasonable evaluation and affirm the certifiable chances of the convicted. Additionally, commonplace stages should be transported off to give primer data on the web and confine media to a colossal degree (Ministry of Electronics & Information Technology, Government of India, 2021). Practical systems for far-reaching trade crusades from the last few pandemics ought to be perceived and executed. These messages should be socially revamped and interpreted in a few vocabularies to develop outreach. For instance, the media joined to talk about the new unbelievable phenomenal respiratory conditions, H1N1 flu, Middle East respiratory issue, Ebola, and H7N9 flu scourges through and through, lessening the absolute spread of these diseases by impelling solid practices. Accurate data concerning COVID-19 can, in like way, be relied upon to have comparable outcomes.

Use Of Telemedicine and Ensuring Transparency, Good Information Governance, and Leadership Across Communities and Institutions

Conversely, telemedicine ought to be utilized to give affirmation-based ideas, and enlightening associations should be made accessible for individuals with restricted acceptance of minds, like individuals in ordinary districts, the old, and people who are completely stuck at home because of lockdown (Abbasi, 2021). This will reduce the bet of clinical focus obtained pollution and decrease the utilization of ventilators and individual wary staff for informed authorities. Straightforwardness, mindblowing data association, and drives across associations and affiliations should be raised to reduce misleading data related to COVID-19, which is fundamental to finally help the masses encounter the success and psychosocial heaps of this pandemic.

Fact-Based and People-Centered COVID-19 Crisis Communication Strategy

The impacts of COVID-19 cut across all socioeconomics. It is troublesome not to frame an assessment on a suffering pandemic that keeps on compromising lives, jobs, and GDP (Su et al., 2021). In any case, given the individual and monetary results attached to one-sided and misdirecting or explicitly bogus and pernicious data, media experts, well-being specialists, and government authorities need to foster a reality-based, individuals-focused COVID-19 emergency correspondence procedure (Su et al., 2021). A reality-based and individual-focused emergency correspondence system is characterized as correspondence that tries to convey realities that make a difference to individuals without outlining the numbers or measurements, considering individual perspectives or ulterior intentions.

Along these lines, benevolent data can be successfully conveyed to people without potentially harmful side effects (Su et al., 2021). It is critical to note that instructive intercessions may likewise be required for medical care experts as a developing assortment of examinations. This shows that medical care experts regularly need fundamental degrees of information or hazard discernment and should have been cautious regarding COVID-19 deception or disinformation (Su et al., 2021). Considering the significant job that medical care experts serve in persistent schooling and how numerous medical services experts likewise face significant emotional wellness challenges, instructive intercessions might be steady, intending to info-demic-instigated difficulties these forefront laborers face.

In conclusion, spreading rumors, hoaxes, and misinformation is an epidemic that characterizes the

COVID-19 pandemic. If it is not well-managed, it will have uncontrollable effects on preventing the disease. The information being communicated to people should be valid and credible. The way it is displayed also affects the recipient's mental health or wellness. For instance, the people receiving the information may be shocked and distressed if no hope is offered at the communication's end. Generally, amid worldwide pandemics like COVID-19, emergency correspondence can critically impact lessening apprehension and vulnerability while motivating a bound together battle against well-being dangers. There presently cannot be a public arrangement or one-sided correspondence during a pandemic. However, considering the articulated requirement for important media assets during COVID-19 for everyone's benefit, well-being specialists and media experts must move forward and end info-demic and slanderous attacks. Partners can fight incorrect revealing with tenable, solid, and dependable data close by very much evolved devices and procedures in emergency correspondence. Straightforwardness and authenticity will, at last, assist with saving individuals' well-being and prosperity while taking worldwide media consideration back to a certified general well-being concern: how to forestall COVID-19 from spreading.

Chapter 5: Case Studies on The COVID-19 Management Schemes and Lessons Learnt Per Country (Brazil and Kenya)

Before embarking on the systems that different governments and agencies put to curb the pandemic and the lessons learned from them, it is essential to study the existing relationship between psychological health and economic stability. The economic impact of the COVID-19 pandemic is enormous, and various countries across the globe have applied different approaches to ensure the effect is reduced. While the methods by many governments and agencies were directed toward reducing the pandemic, these approaches had a lot of psychological implications. Since economic impact directly affects the population's mental health, understanding the various techniques used by multiple countries to mitigate the effects of the pandemic in this scenario will be involved in understanding how those approaches impact the psychological well-being of the population. For instance, if the schools are closed, and a lockdown is placed, what are the economic consequences and psychological effects on people?

Impact of Economic Stability on Psychological Health

As per a late-evolved passionate pestilence bend, nations will encounter two pinnacles of adverse emotional wellness outcomes (Ransing et al., 2020). The primary pinnacle is overwhelmed by tension and compares to the top in COVID-19 cases. The second pinnacle of negative emotional wellness results contains Post-Traumatic Stress Disorder (PTSD), despondency, self-destruction, confounded melancholy mourning, and backsliding of those with existing problems compared to the post-pandemic timeframe (Ransing et al., 2020). Given the considerable expected weight of mental issues regarding the COVID-19 pandemic, the psychological well-being reaction will be highly needed.

Any economic crisis presents great chances of secondary mental health effects that can plunge someone into a deep usage of alcohol and other drug abuse (Silva et al., 2018). Notwithstanding, the psychological well-being impacts of the monetary emergency can be balanced by friendly government assistance and other approach measures. For instance, dynamic work market programs pointed toward aiding individuals to hold or recapture occupations offset the psychological well-being impacts of the monetary emergency. Family support programs add to balancing the emotional wellness impacts of the crisis (Ng et al., 2013). Expanding liquor costs and limiting liquor accessibility decrease the destructive consequences on emotional well-being and save lives (Silva et al., 2018). Obligation alleviation projects will assist with decreasing the emotional well-being impacts of the monetary emergency, and open and responsive essential consideration administrations support individuals in danger and forestall psychological well-being impacts. In this sense, it is noteworthy that whenever the economy is unstable, people undergo many straining events that lead to psychological health problems (Ng et al., 2013). Therefore, economic instability can be best described as a cause of secondary mental health problems due to its indirect impact.

Before the COVID-19 crisis, levels and examples of local earnings and outside monetary streams for developing economies were lacking in the Sustainable Development Goals (SDG) (Ng et al., 2013). With unquestionable levels of public commitment and additional pressures provoked by the pandemic on all tremendous wellsprings of progress cash, low-and-focus pay countries may fight to finance their overall prosperity, social, and monetary responses to COVID-19 (Ng et al., 2013). Early discernments feature tremendous commitment and worth outpourings from developing economies that go with a drop in repayments, and bit by bit, this will extend impacts on local cash recently mentioned by the spreading out broad prosperity and monetary crises. How should leaders avoid a progression finance breakdown that would send millions back into desperation and subvert countries' capacity to show up at the SDGs in this troublesome setting? What should be the typical arrangement for a more grounded, wonderful, and practical world where psychological health is not compromised?

The COVID-19 pandemic is affecting the world at a critical moment, especially the developing economies struggling with sustainable development goals and other factors that have made them lag (Di Crosta et al., 2020). The pandemic has now hit, and these countries are forced to realign their development plans to accommodate the pandemic requirements and create space within the program to cater to the pandemic's effects on people's mental well-being. For instance, some are still thinking of the best fiscal policies that would help sustain their population and development agenda while uplifting the most significant chunk of their population living below the poverty level (Di Crosta et al., 2020).

Consequently, when the economy is unstable, the financial flow is poor, and many people are being laid off from their work. Many people find it hard to sustain their daily basic needs, such as food, a healthy lifestyle, and even shelter away from their jobs. Whenever one misses these basic needs, they are depressed and worried about their life. These worries and stressful episodes lead to more psychological discomfort (Di Crosta et al., 2020). These victims often fall for bad habits, such as drinking alcohol and using hard drugs to solve their problems. The worst happens when they enter a double tragedy. They will have to deal with the created drug addiction, and at the same time, they will still have to think of a way to manage the created effects of the poor economy. There is a direct financial effect from lives lost in an episode past human misfortune. Families and friends lose that pay and their in-kind commitments to family pay, like childcare. Of note, the dissemination of COVID-19 fatalities slants old, which implies that a significant number of those probably going to die are done working and are more averse to being the essential supplier for their families (Di Crosta et al., 2020). Experts must remember, however, that people work until a later age in some low- and center-pay nations. Though less inclined to die from COVID-19, many working-age grown-ups still become sick, and their families will feel the monetary weight as they miss work for days or weeks.

Economic instability is created by the three most natural aversion behaviors that are almost evident in all countries. First, it results when the government forces prohibitions on exercises, such as when the public authority orders manufacturing plants to close or close most shops throughout the country. Second, it results when firms and organizations (counting tuition-based schools and privately owned businesses) go to proactive lengths to keep away from disease. Business terminations, whether through government boycotts or business choices, often bring about lost wages for laborers, particularly in the informal economy where there is no paid leave. Finally, it occurs when people decrease movements to shopping, traveling, going out, and other social exercises.

These activities influence all areas of the economy—the well-being area, assembling, retail and different administrations, exchange and transportation, instruction, and others (Ng et al., 2013). Activities like this convert into scaled-down pay through the stock side (discounted creation drives up costs for buyers) and the exciting side (decreased interest from customers harms entrepreneurs and their representatives). These transient monetary effects can convert into decreases in long-haul development. As the well-being area absorbs more assets and individuals diminish social exercises, nations have a less solid foundation (Ng et al., 2013). As schools close, understudies lose chances to learn (ideally momentarily).

Nevertheless, weak understudies may not get back to the instruction framework, meaning lower long-haul procuring directions for themselves and their families and considerably diminishing human resources for their economies. For instance, spontaneous pregnancies rose forcefully in Kenya during the pandemic, probably to some degree an aftereffect of school terminations (Partridge-Hicks, 2020). Over a time of 90 days in lockdown because of COVID-19, 152,000 young Kenyan ladies became pregnant—a 40% increase in the nation's month-to-month normal (Partridge-Hicks, 2020). From early July, these numbers are the absolute most punctual bits of proof connecting the COVID-19 pandemic to accidental pregnancies. General well-being authorities and ladies' privileges advocate stress that the continuous pandemic postpones a sufficient reaction to a developing sexual regenerative well-being emergency. Juvenile moms are less inclined to get back to school, and their kids will probably have less well-being and instructive speculations. Further, well-being laborers are at the forefront of pandemics, and losing some of them to the infection, particularly in nations where they are now hard to find, can prompt demolishing medical issues in the long haul, such as maternal and baby mortality. These all have neediness suggestions past their compassionate ramifications.

Emotional well-being is a piece of general well-being that is disregarded because of its indissoluble nature. It straightforwardly affects any country's human, social, and financial capital. Emotional wellness is more than not having related mental issues and signs; it is also characterized and affected by asset-supporting prosperity and usefulness. Ostensibly, mental capital is needed for solid family and local area work. Positive emotional wellness is a condition of money wherein the individual understands their capacities, can adapt to life's typical burdens, work gainfully and productively, and add to their local area. With decent psychological well-being, an individual's mental and passionate prosperity becomes adaptable, a key factor required to improve the social abilities to stay strong and assent to distressing and discouraging mental circumstances.

The shock of economic insecurity can significantly influence and undermine the public assistance, schooling, and well-being of areas' spending plans. Similarly, as an individual might be worried about everyday stressors, social order and their nation can also be impacted during financial defeats. Independent of these, better regulative arrangements that ensure social prosperity can help work on the versatility of networks against the recognized financial stressors and thus help alleviate public psychological wellness problems.

These arrangements will help avert high paces of unemployment and other pressure-related economic slump outcomes. Alternately, while financial emergencies might have emotional well-being impacts, psychological well-being issues have progressively huge monetary impacts. Since severe mental problems frequently start in pre-adulthood or youthful adulthood, the deficiency of efficiency can be enduring. Mental issues represent more than 33% of years lived with incapacity. Work changes increase mental and enthusiastic requests (Ng et al., 2013). Non-attendance and withdrawal from the work market have expanded because of pressure, uneasiness, and wretchedness-related issues. In numerous European nations, around 33% of new inability benefits are due to psychological well-being conditions, which is expanding (Ng et al., 2013). In this manner, emotional well-being is a significant financial component. The shift from assembling an information society accentuates the significance of emotional well-being for supporting usefulness. Great populace psychological wellness adds to economic efficiency and success, making it urgent for economic development. The effective recovery of European economies appears to rely upon the emotional health of the populace critically—the accompanying segments diagram how nations can defend and uphold psychological well-being amid an economic slump.

The above analysis indicates that mental health is significantly impacted whenever there is increased instability in the economy. Today, everything globally is financially conditioned, and it is not news that people will be troubled whenever they lack finances. A troubled mind is a stressed mind. Therefore, the COVID-19 pandemic has resulted in extraordinary economic instability, which has indirectly increased the cases of mental illness across the globe. However, various countries have applied different approaches to ensure their citizens are safe and have not plunged into the long-term effect of mental illness due to hard economic times.

Lessons from Kenya

Kenya is among the third-world countries on the eastern continent and has an exciting and educating experience in COVID-19 pandemic management. The Ministry of Health-Kenya (2020) affirmed a coronavirus infection case in Nairobi on March 12, 2020, which was the underlying cause since the flare-up in China in December 2019. The case was a Kenyan resident who ventured out to Nairobi, returning from the U.S. through London, U.K., on March 5, 2020. The National Influenza Center Laboratory affirmed her at the National Public Health Laboratories of the Ministry of Health. The patient was clinically steady and was overseen at the Infectious Diseases Unit at the Kenyatta National Hospital. After every one of these, the Government of Kenya, through the Ministry of Health, kept on fortifying measures to guarantee no further transmission of the illness in Kenya (Ministry of Health-Kenya, 2020). The National Emergency Response Committee on coronavirus readiness and Reaction also kept on giving vital initiatives dealing with the entire government as a way to deal with reacting to this case in the execution of moderation measures.

After the first case, the government started a contact tracing scheme where many people were found, some with local infection roots, while others were traced to the border movement. Even with the surge in the infection rate, the Kenyan government was reluctant to close the borders until it reached a certain point where infections started overwhelming, and the available health facilities could no longer contain the COVID-19 patients. The effects were a lot, and the government was forced to enforce necessary mitigation measures as identified below:

1. Consistently and completely clean up with cleanser and water, or use liquor-based hand sanitizer.
2. Keep a separation of no less than 1 meter (5 feet) between yourself and anyone who is hacking or wheezing (Ministry of Health Kenya, 2020).
3. People with a hack or sniffling should remain at home or maintain a social separation, yet try not to blend in with others in a group (Ministry of Health-Kenya, 2020).
4. Keep up with great respiratory cleanliness by covering your mouth and nose while hacking and sniffling with a cloth or tissue or into a flexed elbow (Ministry of Health-Kenya, 2020).
5. Remain at home, assuming you feel unwell with manifestations like fever, hack, and difficulty relaxing.
6. Suspend every public social affair, gathering, challenging campaign, game, and occasion. Typical community gatherings can go on, given they give cleaning and hand washing.
7. Suspend all entomb-school occasions; however, keep schools open.

8. Public vehicle suppliers were coordinated to give hand sanitizers to their customers and standard cleaning of the vehicles (Ministry of

Health-Kenya, 2020).

9. There was a transitory suspension of jail visits for the following 30 days.

10. Kenyans should not manhandle online media or enjoy spreading deception that can cause dread and frenzy (Ministry of Health-Kenya,

2020).

11. There were travel limitations outside the country, except if totally essential, and there was no movement to infection epicenter nations.

12. There was a scattering of data on a routine.

13. There was mass vaccination.

The main aim of these measures was to ensure that the transmission rate was regulated, as that was the priority. However, these measures placed by the government were quite mean, as they did not consider a very critical factor—a more significant chunk of the population lives below the poverty level, and any movement restrictions will affect their income generation activities. Companies that relied on people working as groups were closed, and workers had a mass layoff. Those left could not produce the maximum amount to support the remaining or the outlaid population.

The cases of COVID-19 are rapidly climbing in sub-Saharan Africa, with stresses raised concerning its capacity to deal with the pandemic. Kenya, a low-middle-income country in eastern Africa, was recorded as being at high risk for importation of COVID-19 close to the beginning of the pandemic in Africa. Likewise, Kenya has been represented as having a low Infectious Disease Vulnerability Index (IVDI), showing its high shortcoming to the episode because of fragile prosperity structures. As of 31st July 2020, Kenya has 21,354 certified examples of COVID-19 and 377 death cases (Jaguga & Kwobah, 2020). A 21-section public emergency response board was set up and is liable for controlling and arranging the COVID-19 response coordinated by the National 2019

Novel Coronavirus Contingency (Readiness and Early Response) Plan (Jaguga & Kwobah, 2020). At any rate, the game plan offers no bearing on a mental well-being response, regardless of orders from the Kenya Mental Health Policy 2015-2030 requiring that passionate prosperity care be given during and after disasters.

The Ministry of Health (MOH), through its Division of Mental Health, has kept on and tried to pass on mental prosperity care during the pandemic. Disastrously, the mental well-being response is occurring against a view of an under-resourced passionate prosperity care structure portrayed by closed-off organizations, a severe insufficiency of enthusiastic health workers, and limited financing. This joined with the shortfall of a formal excited health response plan, is demolishing current undertakings highlighted, diminishing the mental prosperity impact of COVID-19 locally. Given the rising episodes of harmful conduct at home, alcohol use during the COVID-19 pandemic in Kenya, and the meaning of direct procedures in containing the pandemic, a more grounded, passionate well-being response is legitimized.

From the above argument, curbing the associated mental health illnesses brought about by the pandemic in Kenya requires a more profound and sustainable approach. The mitigation measures placed by the government are correct and helpful. However, they are not sustainable. Therefore, five critical procedures for reinforcing the emotional wellness reaction to alleviate the destructive effect of COVID-19 on open psychological well-being in Kenya are as follows:

1. The government should set up a formal emotional well-being reaction plan explicit to the COVID-19 pandemic with a portion of the financing for the reaction.
2. The government should offer to prepare local area well-being laborers and local area well-being volunteers on mental medical aid to empower admittance to help those deprived during the pandemic—an increase of portable well-being to expand admittance to mind will help.
3. Leaders should guarantee legitimate observation of the psychological wellness circumstance during the current pandemic; it is recommended that orderly and ordinary reviews are led to consider checking the emotional well-being effect of COVID-19 in Kenya.
4. Significant versatile specialist organizations in Kenya have overview stages that utilize text informing. These comprise possible means through which emotional well-being reconnaissance could be directed. Such a drive is reasonably attainable given the current associations between the MOH and neighborhood portable specialist organizations.
5. Direct, deliberate, and nonstop instant message studies should be made on the emotional well-being effect of the COVID-19 pandemic to illuminate navigation.

Also, since the virus's origin in Kenya is traced back to foreign countries, it is arguable that this is a global problem and requires a global solution, not a national one. Coordination and international collaboration are critical aspects of solving the problem.

Lessons from Brazil

Unlike most countries across the globe, Brazil was highly affected by the pandemic. The government managing to control the infections was a great initiative and a great source of inspiration. The world has been experiencing the dynamic advancement of the SARS- CoV-2 condition that prompts the sign of the new COVID illness, with phenomenal ramifications for general well-being (He et al., 2020). Since January 2020, the WHO has given perceivability to this globalized general medical condition with the Declaration on Public Health Emergency of International Concern (WHO, 2020). Even though Brazil has had an additional chance to set up, the pandemic has hit the nation. It has been tormenting its populace with a few cases and deaths consistently and forcefully (Department of Informatics of the Unified Health System [DATASUS], 2020). Coronavirus contamination introduces itself, with a critical recurrence of its cases, with convoluted indications and may prompt demise.

Brazil is essential for an enormous gathering of nations in the creating scene, with a significant piece of its populace in a circumstance of social weakness (Guedes et al., 2021). The Brazilian well-being framework (Sistema Único de Saúde-SUS) depends on standards of general consideration, extensive consideration, and value of activities considering the civil rights statute. The assets for adapting to COVID-19 are organized considering these support points (Guedes et al., 2021).

SUS is worked by essential medical care (PHC) as the top level of local area care, directing and coordinating different levels and administrations equipped for serving the vast majority because of the Family Health Strategy (FHS) (Guedes et al., 2021). In any case, an extensive level of the Brazilian domain does not have PHC inclusion or lacks these administrations (Guedes et al., 2021). Likewise, a significant layer of this populace is under troublesome financial conditions, with low admittance to social assurance strategies, showing a deep level of imbalance in the nation (Department of Informatics of the Unified Health System (DATASUS,

2020). In March 2020, when Brazil had fewer than 150 deaths because of the sickness, a solid reaction and proof-based approaches were essential to battle this new infection. However, by August 24th, considering information from the World Health Organization (WHO, 2020), Brazilians had multiple times the chance of dying from COVID-19 when contrasted with the total populace since the start of the pandemic. The following six missteps sum up the expected examples from the administration angle of Brazil.

First, an organized reaction among various regulatory levels and vested parties with solid public stewardship and correspondence systems adjusted to the setting are significant in an episode set. Nonetheless, no shared arrangement has been carried out in Brazil or distributed recently. Since April 2020, two Health

Ministers have surrendered, and the position is empty while a tactical general goes about as a break service (Phillips, 2020). The messages regarding the COVID pandemic are conflictive, and an accusation culture is harming the current public talk. Specialists and the populace are not setting contrasts to battle together against the illness, and the splits among political gatherings are extending.

Second, observation and contact following are essential to control the sickness spread (Vecino-Ortiz et al., 2021). The information ought to be solid and straightforward. The data on quantities of cases and deaths because of COVID-19 had even been excluded for May's specific days. In any case, case discovery through testing is a significant disappointment in the nation, as Thornton (2020) indicated.

Third, case treatment is key to the flare-up reaction and should follow a proof-based methodology. Be that as it may, a few Brazilian specialists and even well-being experts are proposing the utilization of medicines with no affirmed viability. This gives a misguided feeling of security to people in general and, in any event, puts the existence of the populace in danger of unexpected auto-drug consequences.

Fourth, minimized gatherings should be covered in a pandemic because of their higher weakness (Kantamneni, 2020). A higher illness sway in vulnerable ethnic groups is, as of now, detailed. A Brazilian parliamentarian bunch proposed a crisis law to secure customary networks of enslaved people's relatives ("quilombolas") and native populaces. Nevertheless, the government's chief power rejected many pieces of the new draft law.

Fifth, staff well-being ought to be upheld entirely during such unpleasant and phenomenal occasions, and staff should have a real sense of security, both financially and genuinely, to be even more likely to give medical care. On August 4th, the Brazilian government rejected a parliamentarian law specifying monetary remuneration for the well-being of laborers and their families in the event of long-lasting inadequacy or passing because of COVID-19 (Freire, 2020). Due to the tricky connection between the well-being labor force and a few private well-being substances, staff frequently do not try to whine about helpless working conditions or the absence of personal defensive gear. At times, COVID's tainted well-being staff is, in any event, losing their positions and pay since they have been utilized on brief or easygoing premises without government-managed retirement. Besides, the prosperity and honesty of the well-being of staff patients have been underlined after the president settled on a public decision to his devotees to "examine" concentrated consideration units (generally limited admittance). To check assumed beds were occupied by COVID-19 patients (Freire, 2020) inferred that clinics, nearby specialists, and the press were giving bogus data to the general population. Finally, assets should be dispensed quickly and adequately to the unprecedented regulatory levels and well-being specialist co-ops to execute procedures.

Chapter 6: New Variants and Their Impact on the Psychological Safety of People

The COVID-19 pandemic is one of the most significant health crises in modern times. The pandemic has led to the closure of several economic activities, and many people worldwide are still waiting for the day the pandemic will end. The pandemic seems to be here to stay, to their dismay. More hopes were raised when the scientists came up with vaccines to help manage the infections. However, the worst has happened, and many people are still in a dilemma. The variants of the virus being experienced are an issue that is disturbing and unsettling. Many people across the globe believe the virus will one day go, and the pandemic will come to an end. However, with the continued expression of various virus variants, it seems the wait will be long, and people will be plunged deep into a state of psychological discomfort. This section will analyze the variants and their impact on people's psychological conditions.

The manifestations of the variants bring about the following questions: Are people more at risk of getting sick? Will the COVID-19 vaccines still work? Are there new or different things one should do now to stay safe? These questions create a mental disturbance, and because of this, the chances that mental illnesses will still increase are high.

Why Does the Coronavirus Change or Mutate?

Geographic detachment will, in general, bring about hereditarily unmistakable variations. Variations of viruses happen when there is a change—or transformation—to the qualities of the virus (Cleveland Clinic, 2021). RNA viruses—for example, COVID— is the epitome of developing and changing continuously. Mutations in viruses—including the COVID causing the

The COVID-19 pandemic—is neither new nor unforeseen (Cleveland Clinic, 2021). All RNA infections transform after some time, some more than others. For instance, seasonal illnesses frequently change, which is why specialists suggest that people consistently get another influenza immunization (UC Davis Health, 2022). While the possibility of an infection changing may sound alarming, it is entirely normal. Infections transform continually. This is particularly valid for diseases containing RNA as their hereditary material, such as COVIDs and flu infections.

All viruses are comprised of a heap of hereditary material (either DNA or RNA) covered by a defensive covering of proteins (UC Davis Health, 2022). When a virus gets into one's body—generally through the mouth or nose—it locks onto one of the cells. The virus's DNA or RNA then, at that point, enters the cell, where it can make duplicates of itself that go off and taint different cells (UC Davis Health, 2022). Assuming the infection can duplicate itself and commandeer enough of one's cells without being cleared out by their safe framework, that is how someone becomes ill. Sometimes, a blunder happens during the infection's duplicating interaction. That is a change. Changes are often minuscule and do not essentially influence how the infection functions or make the virus more fragile (UC Davis Health, 2022). However, sporadically, a change assists the virus with replicating itself or getting into one's cells even more without any problem. Assuming these beneficial hereditary mix-ups are incorporated when the infection repeats, they are passed on and become a piece of the infection's ordinary genome (Lu et al., 2021). Experts can see these transformations gather over the long run, and that is how people get new variations of an infection strain.

These mutations are developing characteristics that enable them to survive in conditions better than their predecessors (Lu et al., 2021). For instance, the *delta variant* that was recently realized presents a significant threat to humanity due to its ease of transfer from one person to the other. Since September 2021, this variant has proven to be the most lethal one and has dominated a larger region. The lethality of this variant makes it a concern, and the more the problem, the more worried people indulge. These trends are expected to increase since people are uncertain about when all this will end (Lu et al., 2021). Uncertainty about the future is a great contributor to mental discomfort. People will always ask what is next for them (Wu et al., 2020)—asking what is next for them without getting the desired or dependable answers plunges many into depression. The side effects of the delta variation seem, by all accounts, to be equivalent to the first form of COVID-19.

Ongoing exploration observed that the delta variation develops even more quickly—and to a lot more superior levels—in the respiratory plot. Notwithstanding, doctors see individuals getting broken down speedier, particularly more youthful individuals. Ordinarily, immunized individuals are asymptomatic or have extremely gentle indications, assuming they contract the delta variation. Their side effects are more like a typical cold, like hack, fever, or migraine, with massive loss of smell expansion.

From this argument, the severity of the delta variant is higher than the initial strain of the virus. Therefore, people must be highly vigilant and be ready to adhere to more strict mitigation measures and preventive measures than before. It is essential to have masks on and follow the preventative measures specified by the CDC and other health agencies, even after vaccination. Despite being thoroughly inoculated to secure against the delta variation and other COVID-19 transformations, the nation's well-being specialists are wearing covers. They are additionally encouraging immunized individuals to keep away from enormous social affairs and cover up inside where the immunization status of others is obscure.

Arguably, another deadly strain called the omicron is more lethal than the other variants. It has mutated to be more dangerous, and its infection ability is relatively high. The omicron variation of COVID-19 is spreading more quickly than different strains of the infection since it has a more significant number of changes than some other variations up to this point (UC Davis Health, 2022). Many transformations are in the spike protein, which connects with how the infection is spread. The exceptionally infectious COVID-19 variation is causing an expansion in cases and an appeal for testing. On the off chance one has been presented to somebody with the infection or has COVID-19 manifestations and is hanging tight for a test or outcomes, they should remain home and segregate from others.

The variations, including delta and omicron, cause comparable COVID-19 side effects, including hack, fever, and weakness (UC Davis Health, 2022). A few proofs show that fewer individuals with omicron lose their taste and smell. Omicron is also more averse to causing severe sickness, such as pneumonia, that might require therapy in the medical clinic. In thoroughly inoculated and supported individuals, omicron side effects will generally be gentle. In unvaccinated individuals, side effects might be very extreme, perhaps prompting hospitalization or even demise. Omicron is additionally bound to cause side effects regardless of whether one has been uncovered or tested positive before (UC Davis Health, 2022). The vast majority who test positive for any variation of COVID-19 experience a few indications for a long time. Individuals with long COVID-19 manifestations can first encounter medical issues for at least a month after contamination, as per the CDC (2021). The new omicron variation BA.2 seems, by all accounts, to be around half more contagious than the first omicron strain BA.1. Starter information recommends omicron BA.1 causes similar seriousness of sickness and indications, yet it influences more youthful individuals (Karim & Karim, 2021).

There is no foggiest idea of how everyday reinfection is, yet there are reports that few individuals have been tainted with omicron BA.1 and, within a month, contaminated with omicron BA.2 (Karim & Karim, 2021). This form of omicron is more irresistible: it can beat antibodies, past contamination invulnerability, or dodge resistance because of its changes (Karim & Karim, 2021). To secure oneself, getting COVID-19 immunizations is yet significant. The second most crucial thing is veiling. Wearing an N95 will give the best insurance against these new variations.

The CDC (2021) indicates that omicron probably spreads more effectively than the first COVID-19 infection. By all accounts, it seems to be significantly more infectious, contrasted with the delta variation, as it immediately turned into the dominant strain in the U.S. (Jansen et al., 2021). The CDC expects individuals to spread omicron regardless of whether they are inoculated or do not have manifestations. Omicron is bound to cause less severe disease contrasted with different variations. For instance, starter information proposes that contamination with omicron brings about hospitalization at a lower portion of the rate determined with the disease with delta (Jansen et al., 2021). Considering how irresistible omicron is, everybody must play it safe, including immunizing and wearing a cover. The more contaminated individuals, the more confident one will see new variations. The well-being specialists' message is that everybody should be immunized against COVID-19. Each time COVID-19 contaminates somebody, there is an opportunity for additional transformations, which can permit the infection to continue to spread. People must remain watchful and get immunized on the off chance that they have not as of now.

How Does the Trend in Variants Affect People Psychologically?

The first wave of the coronavirus pandemic created one of the most crucial psychological discomforts worldwide. Many did not expect the effects to be as observed in the trend. Many are affected by the increased rate of hospitalizations, death cases, job closures, isolations, and unpredicted economic stabilities. However, people's hopes were raised, and they felt okay when the health agencies announced that vaccines had been manufactured to help manage the pandemic. Even after vaccination, the fact that one could still contract the disease is disheartening, especially for those who raised their hopes high. Worst, the trend in the variants is worrying since this is a condition that people must learn how to live with, irrespective of the mental disorderliness it brings about.

The variants bring a feeling of uncertainty about how the world will look or how people will live in confinement due to the ravaging effect of the pandemic. Learning how to cope with these uncertainties is essential lest most people still suffer significant mental health problems. Not knowing is an awkward encounter. As individuals, people are ordinarily inquisitive (Wu et al., 2020). They try to comprehend, anticipate, and control—it assists them in learning and protects them. Vulnerability can feel hazardous because one cannot foresee with complete certainty what will occur (Wu et al., 2020). Subsequently, both hearts and psyches might race. While it is typical to encounter vulnerability as awkward, it is deplorable for some. Clinicians have even proposed that it is hard to adapt to the experience of not knowing (otherwise called bigotry of vulnerability), which could influence psychological well-being close to various conditions. It is not difficult to perceive how vulnerability is connected to emotional wellness. On the off chance that vulnerability can feel hazardous, it may yet care for one's concern and uneasiness. Additionally, if disposing of that sensation of openness feels fundamental, the impulse to clean up repeatedly to ensure one is spotless and safe may likewise feel fundamental. Furthermore, assuming people eventually feel incapable of adapting to the change and capriciousness life tosses at them, then, at that point, it is reasonable that they are in danger of feeling crushed and discouraged.

Experts are not where they can unhesitatingly clarify which job reactions to vulnerability plays in psychological wellness, yet they can circumspectly offer a few prospects because of the examination in general (Wu et al., 2020). While the discoveries are blended, the best proof that narrow-mindedness or vulnerability might cause emotional wellness challenges is nervousness. Various investigations have observed that it might cause or expand indications of anxiety. That is because when one battle to adapt to the experience of vulnerability, their brain might stress and concoct an expanding number of terrifying potential outcomes (Cohn et al., 2022). Some confirmation recommends that people see that their demeanor is more lamentable when they feel less prepared to adjust to the unseen world. The fight with uncertainty may similarly help people not feel awful.

Nonetheless, a low demeanor is only significant for the experience of distress, so full assessment is required (Cohn et al., 2022). Perhaps amazingly, there is minimal confirmation to help the likelihood that difficulty in overseeing weakness has an impact on causing motivations and obsessions. In any case, of the problems that have been examined, this is similarly the locale with the slightest investigation.

Getting what supports psychological wellness hardships is significant because it can assist people with seeing how to offer better help for many who have these encounters (Cohn et al., 2022). Emotional wellness hardships are every day—indeed, they regularly happen together. This brings up the issue: would experts say they are genuinely independent? Over late years, analysts have begun proposing that what supports one psychological well-being trouble may entirely be shared by others (UNICEF, 2021). There is some help for this idea. For example, the most common way of reasoning over and again, pointlessly, about our interests might prompt both uneasiness and melancholy. Therefore, while these hardships appear to be unique on a superficial level, similar cycles might be working. This is an exciting thought. Rather than having innumerable medicines, people could have supported themselves by turning away from these cyclical thoughts, which is helpful for a broad scope of issues. Above all, people should be sure what their regular cycles are—and genuine work is needed to improve.

Our capacity to endure the vulnerability that life gives is one interaction that may be shared across various emotional wellness hardships (Nast, 2021). If this is true, then, at that point, this comprehension could supportively add to the remedial tool stash across various challenges— a likelihood that is now being investigated. For instance, a social, and mental treatment that decreases narrowmindedness or vulnerability may assist with working on a scope of emotional well-being issues. Additionally, narrow-mindedness of exposure may likewise assume a more extensive part: for example, in dietary problems and psychosis. In any case, in the present moment, there is much mystery and insufficient proof straightforwardly testing these thoughts. Eventually, individuals should be upheld to roll out the improvements that will help them the most. Thus, experts want to research and show what those areas of progress ought to be. After a few exciting starting examinations on the connections between vulnerability and psychological well-being, this region is worth sorting out. Up to that point, all have some susceptibility to bear.

Likewise, with numerous parts of this pandemic, complex illustrations are educational. Achievement now and in the future requires an all-of-government approach, including the exploration and disclosure mastery of the National Institutes of Health, the improved ability of the Biomedical Advanced Research and Development Authority, the strategies aptitude of the Department of Defense, and the mechanical skill of the CDC and other general well-being organizations. Cooperation and correspondence should be across all degrees of government. Extraordinary vaccines can be created, assessed, delivered, and conveyed; however, they are compelling when directed. Execution requires state and neighborhood general well-being, local areas, and religious associations, and believed nearby couriers must participate, notwithstanding different government and state accomplices and assets. Fair access should be highly needed, worked with at each progression, and constantly checked. For instance, the beginning of COVID-19, antibody organizations depended on advanced booking apparatuses to extend and speed up admittance to immunizations. A few grown-ups had restricted, or no admittance to advanced instruments, and networks quickly reacted through neighborhood endeavors to plan arrangements for these people by phone or in-person contact.

One versatile achievement of the COVID-19 antibody execution is the extension of general well-being (Nast, 2021), like the CDC's v-safe (a program new for COVID-19 that has enlisted more than expected—10 million inoculated members). Antibodies were given in drug stores, public venues, miniature and enormous parking garages, and portable centers in rustic and metropolitan networks. The pandemic has demonstrated that practically any area can quickly be changed over to give clinical local area care, like testing and immunization. Creative computerized stages provide novel approaches to public investment in immunization security surveys and impart the enormous advantages of little possible dangers to guarantee to proceed with trust. In a generally decentralized medical services framework, connecting electronic information from individuals and across various settings can be tested; endeavors in information modernization consider coordinating information frameworks to catch deidentified news broadly and share that comprehensive information freely (Nast, 2021).

While disinformation about COVID-19 antibodies remains scattered, more than an expected 75% of U.S. occupants five years and more have started immunization in the previous year (Cohn et al., 2022). This leaves around 25% of unvaccinated qualified people in the US, most by decision (Cohn et al., 2022) like wearing a head protector while riding a bicycle or wearing a safety belt while in a vehicle, getting regularly suggested antibodies is a generally believed anticipation measure and a central general well-being mediation. Throughout recent years, with people depleted from the pandemic's individual, financial, and mental costs, disinformation about COVID-19 immunizations quickly invaded networks (Cohn et al., 2022). Even though spreading antibody reluctance at the populace level is simple, balancing disinformation is tricky regarding web-based media's far and comprehensive utilization. It frequently requires one-on-one conversations, undivided attention, regard for individual worries, and insightful solutions to questions.

Through the COVID-19 vaccination program, the CDC is resolved to complete straightforwardness in the dynamic interaction (Cohn et al., 2022). The informing set assumptions, for example, are automatic objectives, the circumstance of antibody accessibility, and the rhythm of security and viability revealing. While first messages have extraordinary staying power, proof-based changes in direction are unavoidable when new information should illuminate subsequent stages. Conveying these changes that expand on the underpinning of public trust and trust in the emotional cycle is basic. Today, the U.S. is in a better place than one year prior (Cohn et al., 2022). Presently, it is feasible to declare that COVID-19 is generally an immunization-preventable illness.

Simultaneously, antibodies will probably be accessible for youngsters six months and older, broadening assurance for all individuals in the U.S. (Cohn et al., 2022). Prophylaxis for immunocompromised people is currently accessible, and oral antiviral medicines may before long be accessible for those in danger of enormous infection, which will diminish the weight of an overburdened medical care framework. Proposals for supporter dosages because of melting away invulnerability might give more extensive security against arising SARS-CoV-2 variations.

Therefore, the concerned agencies need to address the uncertainties and hope that everything will be right, no matter what is ongoing. The people are supposed to strictly adhere to the set-aside safety protocols for their excellence. Increasingly worrying about the trends of variants will not bring any solution but will increase the dangers of having mental problems. There is much uncertainty now, but people must continue doing what they do to help manage the situation. Fear of tomorrow brings anxiety, which further affects the general operations of people.

Chapter 7: Leadership Skills Needed to Address the New Normal

Abnormal situations demand changed styles of leadership. Using a leadership style applied during the pre-COVID period to try and address the challenges posed by the pandemic may not give a better result, and the effects may be way too much. While the current times are characterized by depression, PTSD, and other mental illnesses, it is essential to develop a leadership style that considers all these factors (Tienne et al., 2020). People should feel cared for, and compassionate leadership should be highly embraced. It takes a lot of energy and patience to deal with a mentally disturbed individual, but that does not mean it is impossible to do it. The COVID-19 pandemic has created a leadership loophole that must be filled if the world continues and people continue with their everyday operations.

As people rise out of this emergency, it could be enticing to cut spending on leadership improvement with the end goal of adjusting the books and regaining funding. More modest financial plans need not prompt leadership quality to be forfeited with the right concentration. Giving up our grasp of how improvement has been sought after previously, people can accept the truth of the present and what is to come (Duggal, 2019). Reevaluating improvement will mean being available to various mechanical arrangements and bearing the cost of delicate abilities and the significance they merit (Duggal, 2019). Wandering into this new world, leader and individual instructing will assume a huge part, with recent fads in leader training arising to satisfy the adjusted need.

Depression is the primary source of inability worldwide (Chambers et al., 2015). One out of five Americans is influenced by emotional wellness issues, with melancholy being the most widely recognized issue. Diagnoses of depression are increasing quicker for 20- to 30-year-old youngsters than for some other ages. Overall, the problem is assessed to cost $44 billion per year in lost usefulness in the U.S. alone (America's Health Rankings, 2021). However, notwithstanding this colossal and developing cost, numerous businesses adopt a specially appointed strategy to care for sadness among workers. When examining why a colleague performs ineffectively, multiple administrators become mindful of emotional well-being issues. An ideal situation would be if workers felt enabled to report a psychological wellness issue and request a sensible convenience. Their administrator can mediate to limit the harm to the association and help the representatives return as fast as conceivable to total well-being.

Addressing mental health issues resulting from the COVID-19 pandemic requires one to understand and have at their fingertips better communication skills, empathy, hybridized leadership teams, technology savviness, and durability. These will help the leader make better and more favorable decisions to help the concerned people manage their situations without any problems. These identified efforts create a greater need to address the issues without any problem and give people hope that new normalcy will be achieved and life will be better, irrespective of the abnormal times.

Communicate with Honesty

On occasions such as these, the two preconditions for business and the range of sensible business results depend on a quick change. This pushes correspondence to the bleeding edge of functional exercises, and experts are expecting both to stay up with the latest. Leaders are called upon to discuss various purposes and interest groups in inner and outer interchanges. The solution to the overwhelming whirlwind of information and shifting conditions is straightforward: clear, authentic, steady, solid, and reality-based data (Anand et al., 2021). When uneasiness and pessimism pose a potential threat, there is a lousy situation for deluding misleading statements or untrustworthy idealism. It is considerably less of a shortfall of data by and large. This requires a specific serenity, mindfulness, receptiveness to criticism, and a consistent progression of correspondence, both on and disconnected.

Whenever a leader identifies a problem that is judged to be mental by an employee or anybody under their command, it is essential to be compassionate and change the way they communicate. The way one addresses an issue with a mentally disturbed individual will determine that individual's ability to manage the mental illness and make life meaningful (WHO, 2021). It is unfortunate that despite the calls for effective and sober communication from leaders, some still use harsh words and phrases, which demoralize and make the individuals less concerned (Tienne et al., 2020). It makes the person feel pessimistic and increases their chances of losing it and becoming a useless society member.

It is not difficult to believe that a representative with mental illness will initially talk with HR staff about work facilities. However, when you are a leader, your colleague (or one of their associates) will undoubtedly speak with you first (McKinsey & Company, 2020). Since a worker might come to you suddenly, you need to get ready early and find out about discouragement and its manifestations. These include loss of premium, diminished energy, low confidence or control sensations, upset rest, and helpless fixation. If you comprehend the expressions of sorrow, you will expect work execution issues and the sorts of facilities a worker may ask for.

Embrace Compassion and Empathy Through Genuine Relationships

Stir excitement for day-by-day business among the fatigued since a long time ago, and detached workers require affectability and passionate insight (Duggal, 2019). The coronavirus has left individuals in a condition of pain—metaphorically, if not in a real sense—and they need human association and comprehension. Pioneers should stay cautious of the more drawn-out term effect of lockdown, be this any psychological wellness that might have emerged or a deficiency of enthusiasm for the work (Tienne et al., 2020). Showing representatives sympathy might be just as simple as tuning in with genuine interest. It might include offering consolation or motivation. At times, it might incorporate alluding to workers for guiding or choosing to stop a gathering to oblige staff's requirement for a difference in speed or some outside air (McKinsey & Company, 2020). Pioneers who see how to develop significant associations with their staff are bound to have the option to get everyone excited.

As laborers explore new close-to-home and expert difficulties welcomed by the COVID pandemic, specialists say it is a higher priority than at any time in recent memory for directors to lead with sympathy (Tienne et al., 2020). Compassion, the act of seeing things according to another person's point of view, comes more generally to specific individuals than others. It is difficult to change how one is designed, particularly on the off chance that they have been reimbursed every moment of their lives for acting as they do. In a review for the current month by worldwide staffing firm Robert Half, 34% of U.S. laborers said they feel more worn out working than they did a year prior (Robert Half, 2021). Many said the weakness came from an expanded responsibility. In September, 88% of ranking directors stressed losing their most gifted laborers (Robert Half, 2021).

Having a compassionate manager can go far in assisting workers with having a sense of safety and appreciation (Dougan et al., 2020). Here are a few different ways to turn out to be more compassionate: to start with, act naturally mindful. Three factors influence sympathy and empathy: how one is raised, the strict convictions one prefers, and weather conditions make one a thinker or a feeler. Feelers start by thinking about how people in the group will respond to a circumstance. Arguably, feelers are the initial ones to detect who may feel annoyed or left out. They esteem concordance, so they stress over that occasionally. Thinkers start by thinking about the group, its objectives, and what assists them with accomplishing them (Dougan et al., 2020). They are less centered around individual reactions and amicability and less adept at detecting the spirit's high and low points.

The two styles are significant in filling each other's holes. Pioneers must self-evaluate to improve their comprehension of themselves regarding showing sympathy (Dougan et al., 2020). Pioneers ought to likewise converse with their staff, as turning out to be more sympathetic requires criticism. The directing inquiry ought to be, "Is there anything you need a greater amount of or less of from me?" Workers may say they need additional time with their chief, more assets, or less fussing over. Therefore, a pioneer ought to tune in. It is not sufficient to save a chance to converse with workers. Regularly, pioneers do not tune in—truly tune in—to what others are telling them since they are thinking about what is next on their plan for others to say straightaway. Focus on individuals' nonverbal prompts — their nonverbal communication, manner of speaking, look, and how passion they show up (Dougan et al., 2020). A worker may consent to another assignment with drooped shoulders and a scowl. Therefore, a sympathetic individual would get some information about that, while somebody who is more errand-centered would take the "I will do that" at face value. While it is hard to peruse those signs in a vis-à-vis setting, remote work can make it significantly harder, as discussions happen on Zoom or email.

At last, come at the situation from others' perspectives. The more force supports an individual, the more effectively that individual can lose sympathy. Lower-level laborers realize their prosperity relies upon saying and making the best choice (Dougan et al., 2020). During the pandemic, numerous representatives are acclimating to remote work and maybe shuffling their kids' virtual learning. In addition, they may be grieving the demise of friends and family or focusing on wiped-out family members. Supervisors should think enough about their representatives to comprehend if they need additional time for a venture or cannot go to night-time gatherings. Many individuals feel that things are somewhat wild. At the point when the uneasiness level ascents, to have a sympathetic and successful director helping one through that is a gift. Directors should consider how they deal with things as standard workers rather than supervisors.

Biblical Aspects of Compassionate Leadership

As leaders during these challenging times, it is highly recommended that we follow the traits of Jesus and other biblical teachings on compassion since the world is becoming highly unpredictable. The Bible is full of examples of leadership during hard times, and when we learn through that lens, it becomes easy to apply that and come out strong. Our most noteworthy illustration of empathy in the Bible is our Heavenly Father. God comforts his kin with sympathy. The *New International Version* (2009) said, "But you, Lord, are a compassionate and gracious God, slow to anger, abounding in love and faithfulness" (Psalm 86:15). Multiple times in God's word, we see the picture of God securing the Israelites in the shadow of His wings, like a mother bird ensuring her hatchlings. For instance, the *New International Version* (2009) also said, "Have mercy on me, my God, have mercy on me, for in you I take refuge. I will take refuge in the shadow of your wing until the disaster has passed" (Psalm 57:1). God's compassion is endless and timeless. His empathies are new each day. They never come short (*New International Version*, 2009, Lamentations 3:22-23).

Jesus Christ, the Messiah, exemplified empathy during his service on the planet. One of the most piercing instances of compassion in the Bible is when Jesus showed sympathy at the grave of Lazarus. At the point when Jesus saw Lazarus's companions sobbing, he sobbed close by them (John 11:33-35). Again, our Lord Jesus Christ felt sympathy for individuals, recuperating them and soothing them. He considered the swarm sheep without a shepherd and came to give them reason and asylum. As the Son of God, Jesus had beyond what we can at any point envision. He quit any pretense of everything, intentionally and sacrificially forfeiting himself for us with the goal that we can have everlasting life. We see the best illustration of empathy in the Bible in that penance. It is a well-known fact that the current pandemic has sent the economy into unrest, leaving organizations enormous and few battling for endurance and a labor force living through incredibly testing times. Against the scenery of dubious business sectors and the UN's conjecture worldwide economy withdrawal of 3.2% in 2020 be outfitted with relevant administration abilities.

Hybridized Teams

While facilitated limitations might see numerous representatives return to work, this boundless time of telecommuting will probably have an enduring effect (Beneria et al., 2020). We hope to see more organizations obliging remote work to different degrees and more noteworthy adaptability regarding working hours. Add to this the developing meaning of the gig economy, and unmistakably, overseeing workers will hold onto new difficulties post-COVID-19. To make this work effectively while defending usefulness and proficiency, pioneers should realize how to cultivate solid, durable groups despite the conspicuous hole between colleagues at the workplace and those telecommuting (Beneria et al., 2020). Empowering variables might incorporate an adjusted work environment highlighting more collective workspaces instead of assigned work area space or cycles, rethinking them for virtual and crossover groups. This compares to a new authority style.

An ordinary plan for getting work done for some organizations suggests being in the workplace all day. In any case, a worker experiencing sadness might come to you and request to go into the workplace later in the day. Rest issues are typical in wretchedness and include sleeping late, such as falling or staying unconscious. Assisting a worker with a plan for getting work done is a sensible convenience. It is upheld by research: flexible work hours increment efficiency, the obligation to the association, and maintenance.

Notwithstanding, on the off chance that you permit flexible hours, research proposes two suggestions. To start with, if necessary, set a window of "center hours" or "center days" in which all colleagues should be at the workplace (Beneria et al., 2020). However, individuals managing misery advantage with structure frequently think it is hard to make the structure themselves. You can help by working with this delicately and responsively. Second, do not let workers with despondency quit collaborating with you or other colleagues. Keep watching for evasion concerning your representative. Withdrawal just intensifies the feeling of detachment that discouraged representatives feel. When left alone, individuals with wretchedness are bound to meditate on the adverse consequences of sadness. This further demolishes the circumstance; on the off chance you presume this is going on, step in and check-in. What is critical here is connecting in a powerful and non-critical way. Exploration recommends that social connections at work can go about as support against wretchedness and that more grounded associations with directors and friends can bring down the gloom.

Technology Leveraging

The unpredictable times and conditions that the pandemic has contributed to require a formidable technological approach to manage. Leaders are challenged to adapt to specific technologies that will help address people's challenges. The public health emergency has seen tech stages. For example, Zoom and Microsoft Teams have become the overwhelming focus. Tech arrangements of this sort, alongside those working with work processed by the executives, project-based cooperation, and correspondence overall, will remain pertinent in the aftermath of COVID, not least guaranteeing organizations can withstand future waves and other expected pandemics. Robotic technology, expanded reality, extensive formation, artificial reasoning, and the Internet of Things might end up being extra wellsprings of corporate life span.

Inappropriately utilizing the innovative choices out there and building a solid computerized presence, those chiefs who can draw upon skill in the space of distributed computing, advanced promoting, web improvement, displaying, and coding will stand far superior to the rest. In the interim, the elevated significance of information in dynamic will require new details and media education qualities, critical thinking, and basic reasoning.

Agility And Resilience in Volatility, Uncertainty, Complexity, and Ambiguity-Abounding (VUCA) Times

The COVID-19 emergency has established the embodiment of VUCA conditions: volatility, uncertainty, complexity, and ambiguity abound (Worley & Jules, 2020). Where already more limited-size organizations might have had the option to dispense with certain factors by unmistakably characterizing their degree or fitting, their administrations to a particular area and the sweeping inclusion of COVID-related intricacies leaves no undertaking immaculate.

Under these conditions, the capacity to respond rapidly to change, showing adaptability and agility in consolidating arranged exercises with adjusted ones or shifting directions, is profoundly respected. A triumphant mentality, flexibility, and the premonition to manage the normal and surprising the same, regularly attempting to trim spending plans recently, are famous (Worley & Jules, 2020). With pioneers under tension as per standard procedure, they should have the option to reprioritize immediately without failing to focus on the master plan. They will likewise have to comprehend that, occasionally, the best authority system is to concede to not having every one of the appropriate responses, realizing when to acquire the specialists to settle on, and sure all business choices depend on reliable data and not predisposition.

These times are exceedingly rugged, and leaders must remain agile as they take their followers through better approaches to help them overcome depressing scenarios. Discouraged individuals can be exceptionally self-basic—maybe combatting disappointments centers around supporting and praising snapshots of accomplishment, for example, when workers fulfill time constraints. Additionally, research shows that individuals scrutinized by somebody they see as profoundly incredulous are less ready to enact neurocircuits that control negative feelings. Therefore, inspiration in discouraged representatives encourages them, notwithstanding dangers and discipline (Worley & Jules, 2020). Examination proposes that clarifying the positive need for tasks as an inspiration device is more compelling than sharing the hindering expenses of an incomplete undertaking. Successful leaders outlined tasks as far as advantages, significantly expanding their apparent allure and fortifying inherent inspiration in representatives. Check-in with your representatives consistently and ensure the work tasks match your staff's current capacities and gifts. If your representative keeps failing, you might be enticed to allot to the humble worker's undertakings or punish the worker with outstandingly troublesome errands that power the representative to work more earnestly. It is conceivable that your representative feels like both situations have effectively happened, in all actuality.

Furthermore, know the qualities of your workers, and play to those qualities. Suppose your workers feel like errands are intended for them. In that case, they will be bound to see the assignments as significant, complete them even more rapidly, and experience a feeling of approval. This technique could be both quickly and eventually valuable. Individuals who feel like tasks are helpful and cooked explicitly to their capacities are more intrigued by the functions and experience decreased degrees of discouragement in the long haul.

In conclusion, coronavirus is a philanthropic misfortune that disturbs many lives. Outstanding leadership is vital, and there is maybe one silver lining: this emergency addresses a chance for pioneers to make more group union and development, even with misfortune. Midway set out an unmistakable arrangement of needs and locally executed them with a deep level of chief neighborhood independence. In a powerful circumstance, the organization excessively incorporated dynamic pivots in line with the association's capacity to answer rapidly and successfully. Viable pioneers engage chiefs to settle on the best choices, remembering an unmistakable arrangement of significant business needs, like keeping representatives safe and acting morally toward clients. Clear limits should be set up when a nearby director needs approval from the middle management.

Chapter 8: Poverty, the Coronavirus Pandemic, and Mental Health

The COVID-19 pandemic has aggravated poverty, leading to unimaginable mental health problems worldwide. The fear of what tomorrow has or will offer can be one of the worst nightmares a person can face (Kumar & Kumar, 2020). It is blatantly visible globally that there is an acute rise in mental health diseases or conditions within the countries defined as low- and middle-income countries. These patterns have exacerbated outrageous poverty; however, they have likewise actuated mental pain and infirmity, in this manner fundamentally influencing the general psychological wellness working of individuals across the globe (Knifton & Inglis, 2020). Mental weakness and poverty are firmly connected in a complicated negative cycle (Kumar & Kumar, 2020). It has been observed that except for sicknesses—for example, schizophrenia, where the hereditary component could be a predominant variable—outrageous material hardship coming about because of the absence of productive business, poor nourishment, and absence of basic conveniences could go before dysfunctional behavior. Individuals impacted by material hardship are known to be at a higher gamble of dysfunctional behavior (Knifton & Inglis, 2020). The pandemic has inflicted significant psychological effects on people living below the specified poverty lines, making life even harder than before (Kumar & Kumar, 2020). The cost of living has exponentially increased, and more funds are channeled into taking care of the pandemic and its effects instead of other social solutions. This chapter will analyze the correlation between the pandemic, poverty lines, and mental health effects. This analysis will quickly determine what approaches the concerned agencies can apply to reduce the pandemic effects on the communities living below the poverty line. Addressing the effects of the pandemic on the mental health of people living below the poverty line will be an effective way to help create policies that would help minimize the effects.

Poverty and Mental Health

World Bank defines extreme poverty as living a life where one earns below $1.90 daily (Kumar & Kumar, 2020). Even before the pandemic struck, many people were living below this line. Their lives are characterized by hunger, overcrowded living conditions, poor accessibility to healthcare services, limited opportunities regarding educational and economic aspects, and extreme life events brought about by being socially vulnerable. These conditions are prerequisites to increased mental health distress and illnesses. Since one will have no better access to the necessities of life, they will end up being worried and, in the process, unable to think straight. Over the long haul, mindfulness in emotional wellness perceives that this endless loop of poverty and chronic mental sickness should be seen according to everyday freedoms.

The emphasis on neediness easing inside well-being sciences needs a superior arrangement with the worldwide improvement plan (Patel et al., 2018; Kumar & Kumar, 2020). Individuals living in poverty likewise experience lower personal satisfaction and show low prosperity pointers, like chronic frailty, social and actual security, and political opportunity and decisions. Destitution builds the gamble of psychological maladjustments that improve the probability of financial slump at individual and family levels (Patel et al., 2018). People with psychological maladjustment are defamed and effectively victimized, and their capacity to hold livelihoods and access potential open doors is diminished. The expense related to emotional wellness medicines likewise increases the monetary weight on the individual and the family (Patel et al., 2018). This shows that living with psychological sickness can make one more unfortunate over the long run.

The emotional wellness of people is molded by the social, ecological, and monetary circumstances where they are conceived, developed, worked, and aged. Destitution and hardship are critical determinants of kids' social and conduct improvement and grown-up psychological well-being (Kumar & Kumar, 2020). People living in the most denied regions report more elevated levels of mental weakness and lower levels of prosperity than those living in the most princely regions. In 2018, for instance, 23% of men and 26% of ladies living in the most denied areas of Scotland revealed degrees of mental trouble characteristic of a potential mental issue, contrasted with 12% and 16% of people living in all deprived regions (Patel et al., 2018). There is also an undeniable connection between regional hardship and self-destruction in Scotland, with suicides multiple times more probable in the most denied regions.

Emotions imbalance in emotional well-being arises from the get-go throughout everyday life and becomes more articulated throughout youth. In one accomplice study, 7.3% of 4-year-olds in the most denied areas of Glasgow were evaluated by their instructor as showing "strange" social, conduct, and personal hardships, contrasted with just 4.1% in all deprived regions (Knifton & Inglis, 2020). By age 7, the hole between these gatherings had augmented considerably:

14.7% of kids in the most denied regions were appraised as having "strange" hardships, contrasted with 3.6% of kids deprived (Knifton & Inglis, 2020). National information from parental evaluations of kids' conduct showed a comparative example. At around 4 years old, 20% of kids living in the most denied areas of Scotland were evaluated as having "fringe" or "unusual" levels of trouble, contrasted with just 7% residing in all deprived regions (Knifton & Inglis, 2020). These stats mirror a more extensive example of financial imbalances in well-being noticed in uniformity. The essential drivers of these disparities are primary contrasts in economic gatherings' admittance to monetary, social, and political assets, which thus influence well-being through a scope of more prompt natural, mental, and conduct processes. A broad spectrum of chance variables is typical among low-pay bunches, including low degrees of seen control, unfortunate practices like smoking, and low degrees of active work compared to their counterparts. These are best perceived as systems that connect the underlying reasons to the disparity in well-being results.

Whenever individuals get analyses of psychological sickness, their work execution and societal position decline; without a lot of treatment accessible, these people continue experiencing disruption, gradually falling back from their positions, families, and companions (Knifton & Inglis, 2020). These people lose work, which implies they lack pay, wind up without an emotionally supportive network, and bring about neediness. Specifically, ladies with extreme psychological maladjustment or those in lower positions experiencing severe dysfunctional behavior are bound to confront complex destitution (Knifton & Inglis, 2020). Since society regularly peers down on ladies and people of the lower position framework, they are the most drastically averse to getting treatment or help when they analyze psychological instabilities.

On the opposite side of this relationship, neediness, which many depict as an absence of work and pay, bothers psychological maladjustment. Mental issues—for example, sadness or nervousness—can be created and strengthened whenever people do not have the necessities for endurance (Knifton & Inglis, 2020). Without therapy, these problems develop, driving clinical experts to determine that people have serious psychological instabilities. Out of those in neediness, ladies, people of the lower stations, and people with SMIs experience the most, as they have the most challenging time tracking down work or getting outside help. Conversely, untreated psychological sicknesses can prompt, or further fuel neediness, yet uncontrolled destitution can cause dysfunctional behaviors, too, making this connection between emotional wellness and neediness.

Both neediness and imbalance are harmful to people's psychological well-being, bringing about more significant levels of dysfunctional behavior among the least fortunate, most avoided, and most underestimated individuals in nations across the world. The argument that poverty is firmly associated with more unfortunate psychological well-being in social orders worldwide is presently broadly settled in research. A report composed by Ridley et al. (2020) and distributed by the U.S.

National Bureau of Economic Research tried to look behind the measurements and investigate why this occurs and how the connections between neediness and psychological sickness can be broken. It drew on proof from the Global South and more affluent countries to recognize what drives the two-way link between poverty and poor psychological well-being and what mediations could be utilized to break its hold on individuals' lives. Inputting things for the survey made a distinct warning from the existing examination.

Economic development is probably not going to work on emotional wellness. Similarly, it works on actual well-being. While numerous parts of well-being connect emphatically with GDP per capita at the national level, psychological complications do not. That implies that wealthy nations getting more extravagant psychological wellness does not result from improving their GDP. Ridley et al. (2020) noted that more rich nations have more elevated levels of psychological behavior. So, countries cannot further develop their residents' psychological wellness by developing their economies and becoming richer generally. It cannot be taken at risk. Ridley et al. (2020) further contended that poverty, in many structures, is an immediate reason for psychological sickness. Routine examinations prove this relationship is causal and not driven by precluded factors. For instance, from assessing longitudinal examinations, Ridley et al. (2020) stated that employment misfortune and pay declines—drivers of neediness—regularly go before episodes of psychological maladjustment. They proceeded to investigate proof supporting this affirmation by seeing what happens when individuals move out of neediness, observing evidence that payor abundance increments can work on psychological wellness and that the advantages of against destitution programs are supported past the immediate change in pay.

The most convincing causal proof that destitution causes psychological sickness comes from randomized controlled preliminaries (RCTs) that assessed hostility to neediness programs (Ridley et al., 2020). A few investigations considered how cash move and more extensive enemy of neediness programs affect psychological well-being, including throughout lengthy timelines when the effects of any underlying celebratory responses among beneficiaries were probably going to have worn off. Likewise, the creators noted that poor psychological well-being could be an aftereffect of the dread and truth of poverty. The expectation of financial shocks, in addition to their event, may cause psychological maladjustment. Individuals living in neediness face significant vulnerability and pay unpredictability, and shuffle what are, in effect, complex financial portfolios frequently without admittance to formal protection.

Supported long-run openness to push from dealing with this unpredictability might compromise emotional wellness.

In addition, Ridley et al. (2020) investigated why destitution drives psychological sickness and the pressure of being in need. Recognizing the higher probability of living in contaminated conditions, in unfortunate lodging, and being presented to excessively high temperatures emerging from worldwide warming as "aggravations" to psychological well-being, they can cause maladjustments like lack of sleep (Knifton & Inglis, 2020). Some of these elements, particularly contamination and congestion, have likewise been distinguished as reasons for the more significant COVID-19 disease and mortality among individuals from minimized and racialized bunches in the U.K. (PHE, 2020).

While investigating the proof of the significance of poverty from the get-go in life as a driver of later psychological maladjustment, Ridley et al. (2020) contended that openness to destitution right off the bat can undermine psychological well-being in later years. Such effects can be produced in utero by presenting pregnant ladies lacking healthy sustenance or having stress. Destitution may likewise excessively open youngsters to unfriendly shocks while their cerebrums are exceptionally plastic and accordingly significantly sway mental health, mental capacity, and psychological well-being in immaturity and adulthood. Openness to wrongdoing, particularly rough wrongdoing, was additionally investigated as a system, expanding the gamble of dysfunctional behavior among needy individuals. It notes that neediness improves an individual's probability of encountering injuries, including the early passing of a relative and homegrown maltreatment. Also, it observes that pay moves can diminish the gamble of viciousness at home.

In contrast to most states of being, wretchedness and uneasiness also straightforwardly affect how individuals think. Despondency and tension might have financial effects because they reduce people's work capacity. Misery might make individuals have reduced confidence in their abilities, while apprehension might make somebody more disciplined in the work market. Ridley et al. (2020) indicated that psychological maladjustment could expand an individual's chance of poverty, for instance, by catching consideration, causing over-the-top rumination, and misshaping individuals' recollections and convictions about their capacities. Ridley et al. (2020) also noted that ladies experience more significant sorrow levels, which might be connected to relative imbalance and command over family funds. They draw on proof that better parental psychological well-being may work on their youngsters' future enthusiasm and monetary prosperity.

Rise in Mental Health Issues Due to COVID-19

Coronavirus is just one indication of the broken connection between people and nature, which is appeared through environmental change, loss of biodiversity, territory discontinuity, and taming of wild creatures. In 2016, the United Nations Environment Program warned of the overall expansion in zoonotic pestilences. Around 60% of all irresistible sicknesses in people are zoonotic, and the development of zoonotic infections is firmly connected with the exhausting soundness of environments (Knifton & Inglis, 2020).

There is proof from studies in Wuhan, China, where the immediate lockdown was started to prepare for powerful reactions to COVID-19. These show that the lockdown has seriously affected the psychological wellness of the Chinese public. In a cross-country cross-sectional review, Sun et al. (2020) reported that facilitating the lockdown in the city was expressly connected with an upsetting effect on a sizeable number of members. Measures like quarantine, confinement, and social removal have truly affected individuals' emotional wellness and general existence in China. Specialists have shown up at agreement that these actions have set off a wide assortment of mental issues—for example, alarm turmoil, uneasiness, and melancholy in everybody (Qiu et al., 2020).

The aftereffects of a Chinese report surveying the impact of COVID-19 moderation anticipating emotional wellness drove Wuhan's government authorities to create a diverse reaction. That included shaping mental mediation groups for key populaces that will probably be affected by openness to COVID-19 and its unfriendly gambling factors (Kang et al., 2020). These underlying examinations from China have featured the need to address weak populaces, improve psychosocial and clinical consideration access, increase catastrophe readiness and relief, reinforce well-being frameworks, debacle the executives, and mediate (Qiu et al., 2020).

General well-being crises sway the two people and networks. There are short-and long-haul ramifications for emotional well-being and prosperity, as seen in the post-pestilence periods of Ebola, SARS, and MERS (Pfefferbaum & North, 2020). During plagues, notwithstanding an enormous number of the populace experiencing discouragement and tension, floods in maladaptive high-risk practices expanded, like aggressive behavior at home and relational contentions. Issues encompassing trepidation, frenzy, and separation were accounted for (Usher et al., 2020) and canvassed daily in information and web-based media. Chances to screen psychosocial needs and convey intercession support during direct understanding experiences in clinical practice are abridged in this emergency by the enormous scope of home repression (Pfefferbaum and North, 2020).

It is assessed that the pervasiveness of posttraumatic stress jumble in everyone goes from 4% to 41%, and the predominance of significant wretchedness will increase by 7% after the episode (Torales et al., 2020).

The pandemic might deteriorate existing psychological wellness issues and lead to more cases among kids and young people. This is because of the exceptional mix of the general well-being emergency, social seclusion, disturbance of instruction, local peer connections, fatigue, and monetary downturn. These elements adversely sway prosperity (Golberstein et al., 2020). These concerns are also exacerbated for those with existing and fundamental medical issues because of the mass quarantine (Torales et al., 2020). Considering that, universally, 36 million individuals have been tainted, roughly more than 1 million deaths have been recorded starting on October 7, 2020 (World Health Organization, 2020a). We realize that the psychological well-being requirements of contaminated people and those who have recuperated as of late might be altogether high. The separation and forlornness around the palliative consideration of the people who capitulate to the ailment imply that these groups experience severe loss and horrendous melancholy. In many regions of the planet, family members and relatives are defamed and constrained into isolation. Such measures are becoming fundamental freedoms infringement (World Health Organization, 2020b), and endeavors are expected to maintain privileges and everyday freedoms.

Pandemic Increasing Poverty in Settings UnderResourced

The results of the COVID-19 lockdown and adjustment attempts have exacerbated the recent development of disheartening and uniqueness among and inside nations. Recently, the World Bank said that around 71 million total individuals would be collided with incredible poverty, expecting that the COVID-19 pandemic does not change divergence inside nations or that the country's improvement accumulates similarly to everybody. Sub-Saharan Africa and South Asia are depended upon to be the hardest hit district, with 26 million and 32 million individuals in each space, autonomously, projected to be living on USD 1.90 or less a day (Kumar & Kumar, 2020).

Close by cash-related neediness, and there is additionally complex distress. Families are denied 33% or a more substantial proportion of the ten indications of success, direction, and suspicions for familiar comforts set out by the Sustainable Development Goals. The Multidimensional Poverty Index estimates that 2% (or 10 million) of the total individuals, out of the 39 million poor individuals, experience unbelievable, exceptional complex urgency, and another 7.7% (or 40 million individuals) not reached, however, are exposed to outrageous multifaceted desperation (UNDP, 2020). The signs of complicated despondency are where a family has malnourished part, high youngster mortality, few children have reached six years old, kids do not go to class, clean drinking water is missing, they are dealing with food, and further made sanitization is missing. Sufficient, safe lodging is inadequate. Individuals who encounter unprecedented, assorted downfall are at high risk for COVID-19 and regularly fail to agree with the careful step degrees of social limitation (UNDP, 2020). This assessed projection of advancement in real money-related generally disheartening situations (USD $5.5/day) construes that any movement in taking out over-the-top neediness would be delayed by something like 20 years.

The effect of COVID-19 on the repeat of destitution is self-evident and combined. The world is encountering a decrease in GDP, with some spots in the extent of 4% and 8% in 2020 (International Monetary Fund, 2020). The limitation on individuals' progression decided bottlenecks radiating from the lockdown. In enormous cases, the completion of creation units as a success and flourishing reaction has compromised individuals' restriction to supply work and secure compensation. The effect is apparent in the limited and insignificant poor (not overall USD $1.9 reliably) living in LMICs. For instance, considering COVID-19, sub-Saharan Africa would see a move in individuals living under inconceivable dejection to 50 million (42%) by 2021 against the pre-COVID-19 assessments of 37.8% (Kumar & Kumar, 2020).

Overall, in low- and middle-income countries, the hard-hit areas have a genuine level of loosened-up laborers and restricted acceptance of flourishing associations and social security. Without fitting strategy measures, such laborers risk falling into unfathomable urgency and will encounter difficulties in recuperating positions during the recuperation period. In another COVID-19 energetic thriving impact layout facilitated by the WHO, results showed the pandemic has upset or ended central mental flourishing associations in 93% of nations worldwide while the interest in mental success is expanding (WHO, 2020c). Experts know that COVID-19 is zeroing in on the limitation of general thriving and key associations worldwide. There are stresses that the LMICs will be impacted because interests in great staffing, thriving staff, acceptance to success associations, and affirmation-based rules and their favorable spread and work with execution keep on being tested. Non-COVID-19 patients would persist as much as those requiring genuine clinical intervention. Suggested mediations for

COVID-19 joins (a) the appraisal of the exactness of data, (b) offering cordial help, (c) reducing the shame related to the disease, (d) keeping a typical life while sticking to someplace liberated from all mischief means, and (e) utilizing psychosocial associations, especially online associations when required (Rajkumar, 2020). Most of these reactions are compromised in under-resourced settings where success structures cannot acclimate to preventive and promotive enlightening and backing. The treatment disruption is so monstrous that avoidance and partner help associations cannot be animated as quickly as possible.

Those most impacted in this pandemic are weakened—low ordinary people who, in like way, have hidden away predictable circumstances. This is assuming that they have a spot with weak get-togethers, such as more prepared adventurer laborers, misinterpreted ethnic or racial social gatherings, ladies, pregnant ladies, and people with broken rehearses. In such cases, consent to associations and accessible help is compromised (Pfefferbaum & North, 2020; Kumar & Kumar, 2020). Obliged quarantine and checking agreeable settlements and areas as red zones is an immediate aftereffect of COVID-19 openness borders and human and psychological wellness praises infringement. Restricting individuals without sufficient provision for their regular necessities can aggravate individuals strain and shortcomings (Torales et al., 2020). Strong diseases convey an extended impression of dread toward ailment, shame, and social forbiddance of people and associations, as was seen during HIV, Ebola, and related plagues.

There is existing proof that poor mental prosperity can be a clarification and a consequence of destitution and trouble (Knifton & Inglis, 2020). Everyday emotional well-being issues sway the customary uttermost spans of the poor as the singular's standard working is decreased (Knifton & Inglis, 2020). Testing energetic success conditions likewise diminishes the individual's standard limit because of lower employability. In places of reality, material trouble can additionally cause pain, uneasiness, or repulsive strain. Experts comprehend that endless franticness in various districts of the planet can be intergenerational. Pandemics further restrict the poor from moving out of mind-boggling neediness. Despite how energetic flourishing wrecks 1 billion individuals all over the planet, the world's committees spend just around 2% of their successful financial plans on settling this issue. Also, many people need acceptance for crucial psychosocial and mental medications. The interconnections between franticness and energetic well-being swear off looking at the flourishing and care foundation and excuse the political parts that cause an advancement in both (Knifton & Inglis, 2020).

Due to the worldwide COVID-19 pandemic, numerous nations have encountered financial moves, prompting postponed progress in diminishing destitution, with multiple residents falling into neediness. The quick spread of the infection has extreme repercussions for individuals. The results incorporate decreased open positions, lower compensation, less admittance to medical care help, troubles progressing to remote learning, and food uncertainty. The effect of COVID-19 on neediness has particularly impacted ladies, youth, and exiles.

The effect of COVID-19 on destitution influences family government assistance because of less work, open doors, and lower income, which prompts diminished food security. Joblessness has almost multiplied, contrasted with pre-pandemic rates. Pay laborers' functioning hours and income have been cut, particularly affecting ladies. Most families depended entirely on the pay of their private ventures, yet numerous organizations shut down or experienced essentially diminished income because of lockdown limitations. Food security is a main pressing issue for some families. Most families announced that food deficiency is the most crucial test in the family. With the deficiency of occupations and pay, individuals can scarcely bear the cost of necessities like food, water, and medical care help. Some cannot manage the cost of costly food varieties like vegetables, and others can only bear to devour a couple of meals each day.

Conclusive Remarks from the Author

The psychological effects of the pandemic will continue ravaging globally, and adequate measures are supposed to be taken for the problem to be well addressed. The COVID pandemic has extraordinarily impacted the entire world. Its impacts are acknowledged on the economy of numerous nations as well as on any remaining parts of life. One of those viewpoints is the mental prosperity of individuals. The pandemic has distressed and tormented individuals, and they will quite often find it challenging to deal with the mental impacts.

The book *The Psychological Effects from the COVID-19 Pandemic* is ideal and significant. In these difficult times, many individuals are going through incredible emotional wellness situations that require a more substantial comprehension to be tended to. During the pandemic, many individuals had to carry on with a day-to-day existence they had never known. They are compelled to adjust to conditions that are insane and pretty against the accepted practices or lifestyles. For example, social isolation and wearing covers in broad daylight places are typical influences on many individuals. The sensation of getting contaminated and being isolated is a staggering inclination that no one would wish to encounter.

This book presents a comprehensive analysis of the associated psychological effects of the pandemic on various people in society. People engage professionally at different levels in a community, and they are affected by issues differently. In this capacity, the book presents an analysis of how the pandemic has deeply affected the community. With critical review and comments on all aspects of psychological discomfort and distress, this book presents arguments that can help manage the associated psychological effect of the pandemic. The intention behind composing this book is to understand these perspectives and how to oversee them across the impacted populace. The book makes sense of the connection between the pandemic and the mental prosperity of individuals. This is achieved by creating a contention toward the pandemic's experience, impacts, and the board plans. As an attendant, the creator has direct data on what is expected to address the effect. Through close perusing and investigation of this book, one will comprehend the stuff to deal with the related mental impacts of the pandemic and what is expected to try not to be impacted by these impacts.

All readers are not only challenged to read this book but to apply the principles in real-life situations. There are several examples in the book that will give vital information that can help in the management of the pandemic. Also, many will be prepared and furnished with measures and approaches expected to defeat the effect of the pandemic. Despite open information, the infection is digging in for the long haul. The avoidance and control measures proposed by the concerned organizations may not ensure 100% chances of well-being. Thusly, this book offers powerful information on what individuals expect and how to deal with the psychological effects of the pandemic.

References

Abbasi, J. (2021). COVID-19 Conspiracies and Beyond: How physicians can deal with patients' misinformation. *JAMA*, *325*(3), 208. https://doi.org/10.1001/jama.2020.22018

Abbott, A. (2021). COVID's mental-health toll: How scientists are tracking a surge in depression. *Nature*, *590*(7845), 194–195. https://doi.org/10.1038/d41586-021-00175-z

Ahmed Siddiqui, M. Y., Mushtaq, K., Mohamed, M. F., Al Soub, H., Hussein Mohamedali, M. G., & Yousaf, Z. (2020). "Social media misinformation"—An epidemic within the COVID-19 pandemic. *The American Journal of Tropical Medicine and Hygiene*, *103*(2), 920–921. https://doi.org/10.4269/ajtmh.20-0592

Alonso-Galbán, P., & Alemañy-Castilla, C. (2020). Curbing misinformation and disinformation in the COVID-19 era: A view from Cuba. *MEDICC Review*, *22*(4), 45. https://doi.org/10.37757/mr2020.v22.n2.12

Al-Ramadan, A., Rabab'h, O., Shah, J., & Gharaibeh, A. (2021). Acute and post-acute neurological complications of COVID-19. *Neurology International*, *13*(1), 102–119. https://doi.org/10.3390/neurolint13010010

America's Health Rankings. (2021, January 3). *2021 disparities report.* https://www.americashealthrankings.org/learn/rep orts/2021-disparities-report?utm_source=SEM&utm_medium=google

&utm_campaign=HD_SEM&gclid=Cj0KCQjw-

4SLBhCVARIsACrhWLWupWe5fXdMVczpIO QVp9B3S-
Z1Xs4eGf2YFspANW3v_elKd4TaTxEaAvPUE ALw_wcB Anand, P., Puranik, A., Aravamudan, M.,

Venkatakrishnan, A., & Sundararajan, V. (2020). Author response: SARS-Cov-2 strategically mimics proteolytic activation of human ENaC. *eLife*, *9*, 1–

13. https://doi.org/10.7554/elife.58603.sa2

Anand, V., Verma, L., Aggarwal, A., Nanjundappa, P., & Rai, H. (2021). COVID-19 and psychological distress: Lessons for India. *PLOS ONE*, *16*(8), e0255683.

https://doi.org/10.1371/journal.pone.0255683 Andersen, K. G., Rambaut, A., Lipkin, W. I., Holmes, E. C., & Garry, R. F. (2020). The proximal origin of SARS-Cov-2. *Nature Medicine*, *26*(4), 450–452.

https://doi.org/10.1038/s41591-020-0820-9

Arshad, M. S., Hussain, I., Mahmood, T., Hayat, K.,

Majeed, A., Imran, I., Saeed, H., Iqbal, M. O.,

Uzair, M., Rehman, A. U., Ashraf, W.,

Usman, A., Syed, S. K., Akbar, M., Chaudhry, M. O., Ramzan, B., Islam, M.,

Saleem, M. U., Shakeel, W., … Rasool, M. F.

(2021). A national survey to assess the COVID-19 vaccine-related conspiracy beliefs, acceptability, preference, and willingness to pay among the general population of

Pakistan. *Vaccines*, *9*(7),

720. https://doi.org/10.3390/vaccines9070720

Asnakew, S., Amha, H., & Kassew, T. (2021). Mental health adverse effects of COVID-19 pandemic on health care workers in northwest Ethiopia: A multicenter cross-sectional study. *Neuropsychiatric Disease and Treatment*, *17*, 1375–1384.

https://doi.org/10.2147/ndt.s306300

Bavel, J. J., Baiker, K., Boggio, P. S., Capraro, V., & Cichocka, A. (2020). Using Social and behavioral science to Support COVID-19 pandemic response. *Nature Human Behaviour*, *2*, 460– 471. https://www.nature.com/articles/s41562-

020-0884-z

Beigel, J. K., Tomasheck, K. M., Dodd, L. E.,

Mehta, A. K., Zingman, B. S., & Kalil, A. C.

(2020). Remdesivir for the treatment of COVID-

19 — Final report. *New England Journal of*

Medicine, *383*(19), 1813–

1826. https://doi.org/10.1056/nejmc2022236

Beneria, A., Arnedo, M., Contreras, S., PérezCarrasco, M., Garcia-Ruiz, I., RodríguezCarballeira, M., Raduà, J., & Rius, J. B. (2020).

Impact of simulation-based teamwork training on COVID-19 distress in healthcare professionals. *BMC Medical Education*,

20(1). https://doi.org/10.1186/s12909-020-02427-

4

Beyrer, C. (2021). A pandemic anniversary: 40 years of

HIV/AIDS. *The Lancet*, *397*(10290), 2142–

2143. https://doi.org/10.1016/s0140-

6736(21)01167-3

Bialowolski, P., Weziak-Bialowolska, D., Lee, M. T.,

Chen, Y., VanderWeele, T. J., & McNeely, E. (2021). The role of financial conditions for physical and mental health. Evidence from a longitudinal survey and insurance claims data. *Social Science & Medicine*, *281*,

114041. https://doi.org/10.1016/j.socscimed.2021

.114041

Blake, H., Bermingham, F., Johnson, G., & Tabner, A.

(2020). Mitigating the psychological impact of COVID-19 on healthcare workers: A digital learning package. *International Journal of Environmental Research and Public Health*, *17*(9), 2997.

https://doi.org/10.3390/ijerph17092997

Browning, M. H., Larson, L. R., Sharaievska, I.,

Rigolon, A., McAnirlin, O., Mullenbach, L.,

Cloutier, S., Vu, T. M., Thomsen, J., Reigner, N.,

Metcalf, E. C., D'Antonio, A., Helbich, M., Bratman, G. N., & Alvarez, H. O. (2021). Psychological impacts from COVID-19 among university students: Risk factors across seven states in the United States. *PLOS ONE*, *16*(1), e0245327. https://doi.org/10.1371/journal.pone.0

245327

Busari, S., & Adebayo, B. (2020, March 23). Nigeria records chloroquine poisoning after Trump endorses it for coronavirus treatment. *CNN.*

https://www.cnn.com/2020/03/23/africa/chloroqu ine-trump-nigeria-intl/index.html Cancino-Montecinos, S., Björklund, F., & Lindholm, T.

(2020). A general model of dissonance reduction: Unifying past accounts via an emotion regulation perspective. *Frontiers in*

Psychology, *11*. https://doi.org/10.3389/fpsyg.202

0.540081

Carvalho, T., Lima, T., Melani, V., Mendes, M., Pereira, L., & Marson, F. (2020). The scientific production during 2009 swine flu pandemic and 2019/2020 COVID-19 pandemic.

Pulmonology, *26*(6), 340–

345. https://doi.org/10.1016/j.pulmoe.2020.07.00

9

CDC. (2021, June 2). Different COVID-19 vaccines.

Centers for Disease Control and

Prevention. https://www.cdc.gov/coronavirus/201

9-ncov/vaccines/different-vaccines.html

Chambers, E., Cook, S., Thake, A., Foster, A., Shaw, S., Hutten, R., Parry, G., & Ricketts, T. (2015). The self-management of longer-term depression: Learning from the patient, a qualitative study. *BMC*

Psychiatry, *15*(1). https://doi.org/10.1186/s12888

-015-0550-6

Chen, B., Sun, J., & Feng, Y. (2020). How have COVID-

19 isolation policies affected young people's mental health? – Evidence from Chinese college students. *Frontiers in Psychology*, *11*,

665. https://doi.org/10.3389/fpsyg.2020.01529

Chu, D., Duda, S., Solo, K., Yaacoub, S., &

Schunemann, H. (2020). Physical distancing, face masks, and eye protection to prevent person-toperson transmission of SARS-Cov-2 and COVID-

19: A systematic review and meta-

analysis. *Journal of Vascular Surgery*, *72*(4),

1500. https://doi.org/10.1016/j.jvs.2020.07.040

Cleveland Clinic. (2021, January 25). What does it mean that the coronavirus is mutating? *Health*

Essentials from Cleveland

Clinic. https://health.clevelandclinic.org/whatdoes-it-mean-that-the-coronavirus-is-mutating/ Cohn, A. C., Mahon, B. E., & Walensky, R. P. (2022).

One year of COVID-19 vaccines. *JAMA,327*(2),

119. https://doi.org/10.1001/jama.2021.23962

Corpuz, J. C. (2021). Adapting to the culture of 'new normal': An emerging response to COVID19. *Journal of Public Health*, *43*(2), e344–e345.

https://doi.org/10.1093/pubmed/fdab057

Cowling, B. J., Ali, S. T., Ng, T. W., Tsang, T. K.,

Li, J. C., Fong, M. W., Liao, Q., Kwan, M. Y.,

Lee, S. L., Chiu, S. S., Wu, J. T., Wu, P., &

Leung, G. M. (2020). Impact assessment of nonpharmaceutical interventions against COVID-19 and influenza in Hong Kong: An observational study. *The Lancet Public Health*, *5*(5), 279–288.

https://doi.org/10.1101/2020.03.12.20034660

Crum, A. J., Jamieson, J. P., & Akinola, M. (2020). Optimizing stress: An integrated intervention for regulating stress responses. *Emotion*, *20*(1), 120–

125. https://doi.org/10.1037/emo0000670

Department of Informatics of the Unified Health System

(DATASUS). (2020, May 18). Coronavírus

Brasil. *Coronavírus*

Brasil. https://covid.saude.gov.br/

DeTienne, K. B., Hooley, J. M., Larrocha, C., & Reay, A. (2020, January 15). How to manage an employee with depression. *Harvard Business Review.* https://hbr.org/2020/01/how-to-managean-employee-with-depression

Di Crosta, A., Palumbo, R., Marchetti, D., Ceccato, I., La

Malva, P., Maiella, R., Cipi, M., Roma, P.,

Mammarella, N., Verrocchio, M. C., & Di Domenico, A. (2020). Individual differences, economic stability, and fear of contagion as risk factors for PTSD symptoms in the COVID-19 emergency.

Frontiers in Psychology, 11.

https://doi.org/10.3389/fpsyg.2020.567367

Dougan, C., Philips, S., Hughes, D., & Gardiner, K.

(2020). Compassionate leadership during COVID-19: An ABC approach to the introduction of new medical graduates as Foundation interim year 1s (FiY1s). *BMJ Leader*, *5*(3), 199–202.

https://doi.org/10.1136/leader-2020-000323 Douglas, M., Katikireddi, S. V., Taulbut, M., McKee, M., & McCartney, G. (2020). Mitigating the wider health effects of the COVID-19 pandemic response. *BMJ*, m1557.

https://doi.org/10.1136/bmj.m1557

Duggal, H. S. (2019). Self-management of depression:

Beyond the medical model. *The Permanente*

Journal, 23, 18–

295. https://doi.org/10.7812/tpp/18-295

Eikenberry, S. E., Mancuso, M., Iboi, E., Phan, T., Eikenberry, K., Kuang, Y., Kostelich, E., &

Gumel, A. B. (2020). To mask or not to mask: Modeling the potential for face mask use by the general public to curtail the COVID-19 pandemic. *Infectious Disease Modelling*, 293– 308. https://doi.org/10.1101/2020.04.06.2005562

4

El-Elimat, T., AbuAlSamen, M. M., Almomani, B. A.,

Al-Sawalha, N. A., & Alali, F. Q. (2021). Acceptance and attitudes toward COVID-19 vaccines: A cross-sectional study from Jordan. *PLOS ONE*, *16*(4), e0250555. https://doi.org/10.1371/journal.pone.0 250555

Esterwood, E., & Saeed, S. A. (2020). Past epidemics, natural disasters, COVID19, and mental health: Learning from history as we deal with the present and prepare for the future. *Psychiatric*

Quarterly, *91*(4), 1121–

1133. https://doi.org/10.1007/s11126-020-09808-

4

Fenn, J. (2020). Public health measures (Lockdown) and its psychological impact: A general population study of mental health difficulties during the COVID-19 pandemic and resultant lockdown. *Asian Review of Social Sciences*, *9*(1),

55–57. https://doi.org/10.51983/arss-

2020.9.1.1606

Fisher, K. A., Bloomstone, S. J., Walder, J.,

Crawford, S., Fouayzi, H., & Mazor, K. M. (2020). Attitudes toward a potential SARS-Cov-2 vaccine. *Annals of Internal Medicine*,*173*(12), 964–

973. https://doi.org/10.7326/m20-3569

Freire, D. (2020, August 4). Bolsonaro veta benefício a profissionais de saúde afetados pela COVID-19.

CNN

Brasil. https://www.cnnbrasil.com.br/politica/bols onaro-veta-beneficio-a-profissionais-de-saudeafetados-pela-covid-19/

Galea, S., Merchant, R. M., & Lurie, N. (2020). The mental health consequences of COVID-19 and physical distancing. *JAMA Internal*

Medicine, *180*(6), 817.

https://doi.org/10.1001/jamainternmed.2020.1562 Garcia-Llana, H., Panizo, N., Gandía, L., Orti, M.,

Giménez-Civera, E., Forquet, C., D'Marco, L.,

Puchades, M. J., Sargsyan, M., Sanchís, I., Ribera, C., Marco, M. I., Ferrá, C. M., PérezBaylach, C. M., Bonilla, B., Moncho Francés, F.,

Perez-Bernat, E., Sancho, A., & Górriz, J. L. (2022). COVID-19 vaccination improved psychological distress (Anxiety and depression scores) in chronic kidney disease patients: A prospective study. *Vaccines*, *10*(2),

299. https://doi.org/10.3390/vaccines10020299

Guedes, M. B., De Assis, S. J., Sanchis, G. J.,

Araujo, D. N., Oliveira, A. G., & Lopes, J. M. (2021). COVID-19 in Brazilian cities: Impact of social determinants, coverage, and quality of primary health care. *PLOS ONE*, *16*(9), e0257347. https://doi.org/10.1371/journal.pone.0257347

Haddad, C., Bou Malhab, S., Sacre, H., & Salameh, P.

(2021). Smoking and COVID-19: A scoping

review. *Tobacco Use Insights*, *14*,

1179173X219946.

https://doi.org/10.1177/1179173x21994612

Haseltine, W. A. (2021, April 21). Understanding the neurological and psychological effects of

COVID-19. *Forbes*.

https://www.forbes.com/sites/williamhaseltine/20 21/04/29/understanding-the-neurological-andpsychological-effects-of-covid-

19/?sh=216a50372090

He, F., Deng, y., & Li, W. (2020). Coronavirus disease

2019: What we know? *Journal of Medical*

Virology, *97*(7), 719–

725. https://doi.org/10.1002/jmv.257662020 Jaguga, F., & Kwobah, E. (2020). Mental health response to the COVID-19 pandemic in Kenya: A review.

International Journal of Mental Health

Systems, *14*(1). https://doi.org/10.1186/s13033-

020-00400-8

Jansen, L., Tegomoh, B., Lange, K., Showalter, K., Figliomeni, J., Abdalhamid, B., Iwen, P. C., Fauver, J., Buss, B., & Donahue, M. (2021).

Investigation of a SARS-Cov-2 B.1.1.529

(Omicron) variant cluster — Nebraska,

November–December 2021. *MMWR. Morbidity and Mortality Weekly Report*, *70*(5152), 1782– 1784. https://doi.org/10.15585/mmwr.mm705152e3

Jeon, J., Baruah, G., Sarabadani, S., & Palanica, A. (2020). Identification of risk factors and symptoms of COVID-19: Analysis of biomedical literature and social media data (Preprint). *Journal of Medical Internet Research*, *22*(10).

https://doi.org/10.2196/preprints.20509

Jia, H., Guerin, R. J., Barile, J. P., Okun, A. H.,

McKnight-Eily, L., Blumberg, S. J., Njai, R., & Thompson, W. W. (2021). National and state trends in anxiety and depression severity scores among adults during the COVID-19 pandemic — the United States, 2020–2021. *MMWR. Morbidity and Mortality Weekly Report*, *70*(40), 1427–1432.

https://doi.org/10.15585/mmwr.mm7040e3

Jones, M., Beardmore, A., Biddle, M., Gibson, A.,

Ismail, S. U., McClean, S., & White, J. (2020). Apart but not alone? A cross-sectional study of neighbour support in a significant UK urban area during the COVID-19 lockdown. *Emerald Open Research*, *2*, 37.

https://doi.org/10.35241/emeraldopenres.13731.1 Kang, L., Li, Y., Hu, S., Chen, M., Yang, C., Yang, B. X., Wang, Y., Hu, J., Lai, J., Ma, X.,

Chen, J., Guan, L., Wang, G., Ma, H., & Liu, Z. (2020). The mental health of medical workers in Wuhan, China, was dealing with the 2019 novel coronavirus. *The Lancet Psychiatry*, *7*(3), e14. https://doi.org/10.1016/s2215-

0366(20)30047-x

Kantamneni, N. (2020). The impact of the COVID-19 pandemic on marginalized populations in the

United States: A research agenda. *Journal of Vocational Behavior*, *119*, 103439.

https://doi.org/10.1016/j.jvb.2020.103439

Karim, S. S., & Karim, Q. A. (2021). Omicron SARSCov-2 variant: A new chapter in the COVID-19 pandemic. *The Lancet*, *398*(10317), 2126–2128.

https://doi.org/10.1016/s0140-6736(21)02758-6 Khanal, P., Devkota, N., Dahal, M., Paudel, K., & Joshi, D. (2020). Mental health impacts among health workers during COVID-19 in a low resource setting: A cross-sectional survey from

Nepal. *Globalization and Health*, *16*(1), 1–12. https://doi.org/10.21203/rs.3.rs-40089/v1

Knifton, L., & Inglis, G. (2020). Poverty and mental health: Policy, practice and research implications. *BJPsych Bulletin*, *44*(5), 193–

196. https://doi.org/10.1192/bjb.2020.78

Kugler, G., & Thaler, O. (2021, February 18). Unholy vaccine: why some Israelis are afraid of getting the Covid vaccine.

Plus61JMedia. https://plus61j.net.au/editorspicks/unholy-vaccine-israelis-afraid-gettingcovid-vaccine/

Kumar, M., & Kumar, P. (2020). Impact of pandemic on mental health in lower- and middle-income countries (LMICs). *Global Mental Health*, *7*, e35. https://doi.org/10.1017/gmh.2020.28

Littman, R. J. (2009). The Plague of Athens:

Epidemiology and paleopathology. *Mount Sinai*

Journal of Medicine: A Journal of Translational and Personalized Medicine, *76*(5), 456–467. https://doi.org/10.1002/msj.20137

Lotfi, M., Hamblin, M. R., & Rezaei, N. (2020). COVID19: Transmission, prevention, and potential therapeutic opportunities. *Clinica Chimica Acta*, *508*, 254–266.

https://doi.org/10.1016/j.cca.2020.05.044 Lu, J., Li, B., Deng, A., Li, K., Hu, Y., Li, Z., Xiong, Q.,

Liu, Z., Guo, Q., Zou, L., Zhang, H., Zhang, M.,

Ouyang, F., Su, J., Su, W., Xu, J., Lin, H., Sun, J., Peng, J., … He, J. (2021). Viral infection and transmission in a large, well-traced outbreak caused by the SARS-Cov-2 delta variant. *Genomic Epidemiology*, *1*(1), 1–

17. https://doi.org/10.21203/rs.3.rs-738164/v1 MacDonald, N. E., & World Health Organization

Strategic Advisory Group of Experts (SAGE).

(2015). Vaccine hesitancy: Definition, scope, and determinants. *Vaccine*, *33*(34), 4161–

4164. https://doi.org/10.1016/j.vaccine.2015.04.0

36

Malik, A. A., McFadden, S. M., Elharake, J., & Omer, S. B. (2020). Determinants of COVID-19 vaccine acceptance in the U.S. *EClinicalMedicine*, *26*(100495).

https://doi.org/10.1101/2020.05.22.20110700

Mallapaty, S. (2021). India's DNA COVID vaccine is a world-first – more are coming. *Nature*,

597(7875), 161–

162. https://doi.org/10.1038/d41586-021-02385-x

Manjula Bai, H. (2020). The socio-economic implications of the coronavirus pandemic (COVID-19): A review. *ComFin Research*, *8*(4), 8–17.

https://doi.org/10.34293/commerce.v8i4.3293

Maxmen, A., & Mallapaty, S. (2021). The COVID lableak hypothesis: What scientists do and don't know. *Nature*, *594*(7863), 313–

315. https://doi.org/10.1038/d41586-021-01529-3

McKinsey & Company. (2020, March 16). Leadership in a crisis: How leaders can support their organizations during the COVID-19 crisis and recovery. https://www.mckinsey.com/businessfunctions/organization/our-insights/leadership-ina-crisis

Ministry of Electronics & Information Technology, Government of India. (2021, May 7). Advisory to curb false news / Misinformation on coronavirus | Ministry of electronics and information technology, government of India. *Ministry of*

Electronics and Information Technology,

Government of

India. https://www.meity.gov.in/advisory-curbfalse-news-misinformation-corona-virus

Ministry of Health-Kenya. (2020, March 12). The first case of coronavirus disease was confirmed in

Kenya. *MINISTRY OF HEALTH – REPUBLIC OF KENYA.* https://www.health.go.ke/first-caseof-coronavirus-disease-confirmed-inkenya/#:~:text=Port%20Health%20Services-

,FIRST%20CASE%20OF%20CORONAVIRUS

%20DISEASE%20CONFIRMED%20IN%20KE

NYA,in%20China%20in%20December%202019

Mitze, T., Kosfeld, R., Rode, J., & Wälde, K. (2020). Face masks considerably reduce COVID-19 cases in Germany. *Medical Letter on the CDC & FDA*, 209–238.

https://doi.org/10.1101/2020.06.21.20128181

Mukhtar, S. (2020). Psychological impact of COVID-19 on older adults. *https://www.ncbi.nlm.nih.gov/pmc/articles/PMC7 373678/*, *10*(4), 201–202.

https://www.ncbi.nlm.nih.gov/pmc/articles/PMC7

373678/

Nast, C. (2021, December 1). The uncertainties of the omicron variant. *The New Yorker.*

https://newyorker.com/science/medicaldispatch/the-uncertainties-of-the-omicron-variant Neergaard, L., & Fingerhut, H. (2020, May 27). APNORC poll: Half of Americans would get a COVID-19 vaccine. *AP NEWS.*

https://apnews.com/article/donald-trump-usnews-ap-top-news-politics-virus-outbreakdacdc8bc428dd4df6511bfa259cfec44 Nelson, T., Kagan, N., Chritchlow, C., Hillard, A., & Hsu, A. H. (2020). The danger of misinformation in theCOVID-19 crisis. *Missouri medicine*, *117*(6), 510–512. https://www.ncbi.nlm.nih.gov/pmc/articles/PMC7

721433/

Ng, K. H., Agius, M., & Zaman, R. (2013). The global economic crisis: Effects on mental health and what can be done. *Journal of the Royal Society of*

Medicine, *106*(6), 211–214.

https://doi.org/10.1177/0141076813481770

OECD. (2021, May 12). Tackling the mental health impact of the COVID-19 crisis: An integrated, whole-of-society response. https://www.oecd.org/coronavirus/polic y-responses/tackling-the-mental-health-impactof-the-covid-19-crisis-an-integrated-whole-ofsociety-response-

0ccafa0b/#:~:text=The%20COVID%E2%80%91

19%20crisis%20has,to%20health%20services%2

0%E2%80%93%20fell%20dramatically

Omer, S. B., Yildirim, I., & Forman, H. P. (2020). Herd immunity and implications for SARS-Cov-2 control. *JAMA*, *324*(20),

2095. https://doi.org/10.1001/jama.2020.20892

Padhi, B. K., & Al-Mohaithef, M. (2020). Determinants of COVID-19 vaccine acceptance in Saudi

Arabia: A web-based national survey. *Journal of*

Multidisciplinary Healthcare,*13*, 1657– 1663. https://doi.org/10.1101/2020.05.27.201144

13

Partridge-Hicks, S. (2020, August 19). Rise in teenage pregnancies in Kenya linked to COVID-19 lockdown. *Global*

Citizen. https://www.globalcitizen.org/en/content/ rise-in-teenage-pregnancies-during-kenyalockdown/

Pfefferbaum, B., & North, C. S. (2020). Mental health and the COVID-19 pandemic. *New England Journal of Medicine*, *383*(6), 510–

512. https://doi.org/10.1056/nejmp2008017

Phillips, D. (2020, July 1). Brazil loses second health minister in less than a month as COVID-19 deaths rise. *The*

Guardian. https://www.theguardian.com/world/2 020/may/15/brazil-health-minister-nelson-teichresigns

Piltch-Loeb, R., Harriman, N., Healey, J., Bonetti, M.,

Toffolutti, V., Testa, M. A., Su, M., & Savoia, E. (2021). COVID-19 vaccine concerns safety, effectiveness, and policies in the United States, Canada, Sweden, and Italy among unvaccinated individuals. *Vaccines, 9*(10), 1138.

https://doi.org/10.3390/vaccines9101138

Pineo, R. (2021). Four flu pandemics: Lessons that need to be learned. *Journal of Developing Societies*, *37*(4), 398–448. https://doi.org/10.1177/0169796x211047221

Prescott, H. C., & Girard, T. D. (2020). Recovery from severe COVID-19. *JAMA*, *324*(8), 739. https://doi.org/10.1001/jama.2020.14103

Qiu, J., Shen, B., Zhao, M., Wang, Z., Xie, B., & Xu, Y. (2020). undefined. *General Psychiatry*, *33*(2), e100213. https://doi.org/10.1136/gpsych-2020-100213

Raman, S. (2020, April 2). Public health experts worry about spread of COVID-19 misinformation. *Roll Call.* https://rollcall.com/2020/03/18/publichealth-experts-worry-about-spread-of-covid-19misinformation/

Ransing, R., Adiukwu, F., Pereira-Sanchez, V., Ramalho, R., Orsolini, L., Teixeira, A. L., Gonzalez-Diaz, J. M., Pinto da Costa, M., Soler-Vidal, J., Bytyçi, D. G., El Hayek, S., Larnaout, A., Shalbafan, M., Syarif, Z., Nofal, M., & Kundadak, G. K. (2020). Mental health interventions during the COVID-19 pandemic: A conceptual framework by early career psychiatrists. *Asian Journal of Psychiatry*, *51*, 102085. https://doi.org/10.1016/j.ajp.2020.102085

Ridley, M., Rao, G., Schilbach, F., & Patel, V. (2020). Poverty, depression, and anxiety: Causal evidence and mechanisms. https://doi.org/10.3386/w27157

Robert, H. (2021, April 9). One in three remote workers may quit if required to return to the office full time, Robert half survey finds. *Robert Half.* https://www.roberthalf.ca/en/one-in-three-remoteworkers-may-quit-if-required-to-return-to-theoffice-full-time-robert-half

Rossen, I., Hurlstone, M. J., Dunlop, P. D., & Lawrence, C. (2019). Accepters, fence sitters, or rejecters: Moral profiles of vaccination attitudes. *Social Science & Medicine*, *224*, 23–27.

https://doi.org/10.1016/j.socscimed.2019.01.038

Saladino, V., Algeri, D., & Auriemma, V. (2020). The psychological and social impact of COVID-19: New perspectives of well-being. *Frontiers in Psychology, 11*. https://doi.org/10.3389/fpsyg.2020.577684

Scheidel, W. (2009). Roman well-being and the economic consequences of the 'Antonine plague.' *SSRN Electronic Journal*. https://doi.org/10.2139/ssrn.1442584

Shanafelt, T., Ripp, J., & Trockel, M. (2020). Understanding and addressing sources of anxiety among health care professionals during the COVID-19 pandemic. *JAMA*, *323*(21), 2133. https://doi.org/10.1001/jama.2020.5893

She, J., Jiang, J., Ye, L., Hu, L., Bai, C., & Song, Y. (2020). 2019 novel coronavirus of pneumonia in Wuhan, China: Emerging attack and management strategies. *Clinical and Translational Medicine*, *9*(1). https://doi.org/10.1186/s40169-020-00271-z

Shukla, A. (2020). Psychological impact of COVID-19 lockdown: An online survey from India: Few concerns. *Indian Journal of Psychiatry*, *62*(5), 591. https://doi.org/10.4103/psychiatry.indianjpsychiatry_1012_20

Silva, M., Resurrección, D. M., Antunes, A., Frasquilho, D., & Cardoso, G. (2018). Impact of economic crises on mental health care: A systematic review. *Epidemiology and Psychiatric Sciences*, *29*. https://doi.org/10.1017/s2045796018000641

Singhal, T. (2020). A review of coronavirus disease-2019 (COVID-19). *Indian Journal of Pediatrics*, *87*(4), 281–

286. https://www.ncbi.nlm.nih.gov/pmc/articles/P

MC7090728/

Spoorthy, M. S., Pratapa, S. K., & Mahant, S. (2020). Mental health problems faced by healthcare workers due to the COVID-19 pandemic–A review. *Asian Journal of Psychiatry*, *51*,

102119. https://doi.org/10.1016/j.ajp.2020.10211

9

Strand, H., & Parker, D. (2012). Effects of multidisciplinary models of care for adult predialysis patients with chronic kidney disease: A systematic review. *International Journal of*

Evidence-Based Healthcare, *10*(1), 53–

59. https://doi.org/10.1111/j.1744-

1609.2012.00253.x

Su, Z., McDonnell, D., Wen, J., Kozak, M., Abbas, J.,

Šegalo, S., Li, X., Ahmad, J., Cheshmeh

Angi, A., Cai, Y., Yang, L., & Xiang, Y. (2021). Mental health consequences of COVID-19 media coverage: The need for effective crisis communication practices. *Globalization and Health*, *17*(1). https://doi.org/10.1186/s12992-

020-00654-4

Sun, J., Chi, C., Li, C., Buys, N., & Yin, C. (2020). Clinical characteristics of 92,800 patients with confirmed coronavirus disease (COVID-19) infection: A systematic review and metaanalysis. *SSRN Electronic*

Journal. https://doi.org/10.2139/ssrn.3677878

Tang, B., Bragazzi, N. L., Li, Q., Tang, S., Xiao, Y., & Wu, J. (2020). An updated estimation of the risk of transmission of the novel coronavirus (2019-

nCov). *Infectious Disease Modelling*, *5*, 258–

255. https://doi.org/10.1016/j.idm.2020.02.001

Tartof, S. Y., Slezak, J. M., Fischer, H., Hong, V.,

Ackerson, B. K., Ranasinghe, O. N.,

Frankland, T. B., Ogun, O. A., Zamparo, J. M.,

Gray, S., Valluri, S. R., Pan, K., Angulo, F. J., Jodar, L., & McLaughlin, J. M. (2021).

Effectiveness of mRNA BNT162b2 COVID-19

vaccine up to 6 months in a large integrated health system in the USA: A retrospective cohort study. *The Lancet*, *398*(10309), 1407–1416.

https://doi.org/10.1016/s0140-6736(21)02183-8

Tasnim, S., Hossain, M. M., & Mazumder, H. (2020). Impact of rumors or misinformation on coronavirus disease (COVID-19) in social media. *Journal of Preventive Medicine & Public*

Health, *53*(2020), 171–

174. https://doi.org/10.31235/osf.io/uf3zn

Thornton, J. (2020). COVID-19: Lack of testing in Brazil is a "major failure," says MSF. *BMJ*, m2659. https://doi.org/10.1136/bmj.m2659

Torales, J., O'Higgins, M., Castaldelli-Maia, J. M., &

Ventriglio, A. (2020). The outbreak of COVID19 coronavirus and its impact on global mental health. *International Journal of Social*

Psychiatry, *66*(4), 317–

320. https://doi.org/10.1177/0020764020915212

Troyer, E. A., Kohn, J. N., & Hong, S. (2020). Are we facing a crashing wave of neuropsychiatric sequelae of COVID-19? Neuropsychiatric symptoms and potential immunologic mechanisms. *Brain, Behavior, and Immunity*, *87*,

34–39. https://doi.org/10.1016/j.bbi.2020.04.027

U.S. Food and Drug Administration (FDA). (2020, October 22). FDA approves the first treatment for COVID-19. *U.S. Food and Drug Administration.* https://www.fda.gov/news-events/pressannouncements/fda-approves-first-treatmentcovid-19

UC Davis Health. (2022, January 25). Omicron variant: What we know so far about this COVID-19 strain. *UC Davis Health.* https://health.ucdavis.edu/coronavirus/covid-19information/omicron-variant

UNDP. (2020). Multidimensional Poverty in times of

COVID-

19. https://www.latinamerica.undp.org/content/rbl ac/en/home/blog/2020/pobreza-multidimensionalen-tiempos-del-covid-19.html

UNEP. (2016, February 15). Frontiers 2016: Emerging issues of environmental concern. *UNEP - UN*

Environment

Programme. https://www.unenvironment.org/reso urces/frontiers-2016-emerging-issuesenvironmental-concern

UNICEF. (2021, June 15). "This vaccine will bring back hope for us all."

https://www.unicef.org/supply/vaccine-willbring-back-hope-us-all

Usher, K., Durkin, J., & Bhullar, N. (2020). The COVID-

19 pandemic and mental health impacts. *International Journal of Mental Health Nursing*, *29*(3), 315–318.
https://doi.org/10.1111/inm.12726

Vahey, G. M., McDonald, E., Marshall, K.,

Martin, S. W., Chun, H., Herlihy, R., Tate, J. E.,

Kawasaki, B., Midgley, C. M., Alden, N., Killerby, M. E., & Staples, J. E. (2021). Risk factors for hospitalization among persons with

COVID-19—Colorado. *PLOS ONE,16*(9),

e0256917. https://doi.org/10.1371/journal.pone.0

256917

Varatharaj, A., Thomas, N., Ellul, M., Davies, N. W.,

Pollak, T., Tenorio, E., Sultan, M., Easton, A.,

Breen, G., Zandi, M., Coles, J. P., Manji, H., Al-

Shahi Salman, R., Menon, D., Nicholson, T.,

Benjamin, L., Carson, A., Smith, C.,

Turner, M. R., … Michael, B. (2020). UK-wide surveillance of neurological and neuropsychiatric complications of COVID-19: The first 153 patients. *SSRN Electronic Journal*, *7*(10), 857–

882. https://doi.org/10.2139/ssrn.3601761

Varshney, M., Parel, J. T., Raizada, N., & Sarin, S. K.

(2020). The initial psychological impact of COVID-19 and its correlates in Indian community: An online (FEEL-COVID) survey.

PLOS ONE, *15*(5),

e0233874. https://doi.org/10.1371/journal.pone.0

233874

Vecino-Ortiz, A. I., Villanueva Congote, J., Zapata Bedoya, S., & Cucunuba, Z. M. (2020). Impact of contact tracing on COVID-19 mortality: An impact evaluation using surveillance data from Colombia. *PLOS ONE*, *16*(3).

https://doi.org/10.1101/2020.08.14.20158535

Wallen, J. (2020, February 12). Coronavirus: Indian man 'died by suicide' after becoming convinced he was infected. *The*

Telegraph. https://www.telegraph.co.uk/globalhealth/science-and-disease/coronavirus-indianman-died-suicide-becoming-convinced-infected/

Weida, E. B., Phojanakong, P., Patel, F., & Chilton, M. (2020). Financial health as a measurable social determinant of health. *PLOS ONE*, *15*(5),

e0233359.

https://doi.org/10.1371/journal.pone.0233359

WHO. (2020, February 12). COVID-19 public health emergency of international concern (PHEIC) global research and innovation forum. *World Health Organization.*

https://www.who.int/publications/m/item/covid19-public-health-emergency-of-internationalconcern-(pheic)-global-research-and-innovationforum

WHO. (2021). #HealthyAtHome. *World Health*

Organization. https://who.int/campaigns/connecting-the-worldto-combatcoronavirus/healthyathome/healthyathome--mental-health?gclid=Cj0KCQjw-

4SLBhCVARIsACrhWLV91UAfUyfTZtBxM1fh soFcKwMIvDq4ckIHTpXcI4niMs68CPb7XqsaA hw0EALw_wcB

WHO. (2020, March 26). Origin of SARS-CoV-2. *World*

Health Organization.

https://apps.who.int/iris/bitstream/handle/10665/3

32197/WHO-2019-nCoV-FAQ-Virus_origin-

2020.1-eng.pdf

Wilson, W., Raj, J. P., Rao, S., Ghiya, M.,

Nedungalaparambil, N. M., Mundra, H., & Mathew, R. (2020). Prevalence and predictors of stress, anxiety, and depression among healthcare workers managing COVID-19 pandemic in India: A nationwide observational study. *Indian Journal of Psychological Medicine*, *42*(4), 353–358.

https://doi.org/10.1177/0253717620933992

Worley, C. G., & Jules, C. (2020). COVID-19's uncomfortable revelations about agile and sustainable organizations in a VUCA world. *The Journal of Applied Behavioral Science*,*56*(3),

279–

283. https://doi.org/10.1177/0021886320936263

Wu, D., Yu, L., Yang, T., Cottrell, R., Peng, S., Guo, W., & Jiang, S. (2020). The impacts of uncertainty stress on mental disorders of Chinese college students: Evidence from a nationwide study.

Frontiers in Psychology, *11*.

https://doi.org/10.3389/fpsyg.2020.00243

Zhao, X., Rafiq, A., Hummel, R., Fei, D., &

Merrell, R. C. (2006). Integration of information technology, wireless networks, and personal digital assistants for triage and casualty. *Telemedicine and e-Health*, *12*(4), 466– 474. https://doi.org/10.1089/tmj.2006.12.466

Made in the USA
Monee, IL
02 November 2026

67387602R00063